The Institute of Mathematics
and its Applications
Conference Series

The Institute of Mathematics
and its Applications
Conference Series

Previous volumes in this series were published by
Academic Press to whom all enquiries should be addressed.
Forthcoming volumes will be published by
Oxford University Press throughout the world.

Analysing conflict and its resolution

some mathematical contributions

Based on the proceedings of a conference organized
by the Institute of Mathematics and its Applications on
The Mathematics of Conflict and its Resolution,
held at Churchill College, Cambridge in December 1984

Edited by

P. G. BENNETT
University of Sussex

CLARENDON PRESS · OXFORD · 1987

Oxford University Press, Walton Street, Oxford OX2 6DP

Oxford New York Toronto
Delhi Bombay Calcutta Madras Karachi
Petaling Jaya Singapore Hong Kong Tokyo
Nairobi Dar es Salaam Cape Town
Melbourne Auckland

and associated companies in
Beirut Berlin Ibadan Nicosia

Oxford is a trade mark of Oxford University Press

Published in the United States
by Oxford University Press, New York

British Library Cataloguing in Publication Data
Analysing conflict and its resolutions: some mathematical
contributions based on the proceedings of a conference
organized by the Institute of Mathematics and its
Applications, The Mathematics of Conflict and its
Resolution, held at Churchill College, Cambridge in
December, 1984.——(The Institute of Mathematics
and its Applications conference series)
1. Conflict management——
Mathematical models
i. Bennett, P.G.   ii. Series
303. 6'072   HD42
ISBN 0–19–853611–9

Printed in Great Britain by St Edmundsbury Press,
Bury St Edmunds, Suffolk

"Your enemy is never a villain in his own eyes. Keep this in
mind: it may offer a way to make him your friend.

If not, you can kill him without hatred ... and quickly."

From "The sayings of Lazarus Long" in Robert Heinlein's
<u>Time Enough For Love</u> New English Library, 1974.

CONTENTS

EDITOR'S INTRODUCTION

The papers collected here have their origins in a conference
on "The Mathematics of Conflict and its Resolution", held at
Churchill College, Cambridge in December 1984, under the aegis
of the Institute of Mathematics and its Applications.

Conflict in various forms is to be found all around us, in
international relations, in politics, in economic (and in
biological) competition, in relationships between individual
people. Though the most obvious "conflicts" may be thought of
as wars, revolutions, riots, strikes, and hostilities of other
sorts, conflict in a more general sense is a natural
consequence of our power to take actions that affect others.
Such conflict is seldom absolute, but coexists with the need -
at least at some level - for cooperation. Similarly, conflict
can be studied from many different standpoints. Taking the
international sphere for example, these perspectives range
from that of the military analyst, to that of the researcher
interested in how conflicts arise, develop, and may eventually
be resolved, and again to that of the campaigner for
disarmament. Though these concerns are by no means mutually
exclusive, this conference was a little unusual in including
contributions from across a very broad spectrum. The resulting
variety of viewpoints proved to be most stimulating.

A common element in the contributions lies in the use of
some form of mathematics. For those not already familiar with
the field, it should be said that this does not necessarily
mean high levels of quantification or a supposed finding of
precise "right answers". Sometimes, of course, variables can
be measured precisely, and technical questions answered with
reasonable certainty. More often however, a more important
role for mathematics is in providing structures within which
arguments can be framed and followed to their conclusion. It
is then easier to ensure that assumptions are clearly stated,
to examine the resulting models rigorously, and to explore
the consequences of different hypotheses. That, at least, is
the hope.

It would clearly be impertinent to try to summarise the
various contributions in the space available here. Nevertheless,
the reader may find some rough guidance helpful. The book is
arranged in three sections (though all are interconnected):

"Analysing Decisions in Conflict", "Arms Races and
International Systems" and "Modelling Battle". Within each,
the contents are ordered so that - as far as possible - papers
providing a reasonably general overview of the ideas involved
appear first, followed by those dealing with specific problems
or more advanced theoretical developments. While the second
and final groupings deal with specific areas of study, the
first deals with a set of approaches that can in principle be
applied to conflicts in any field, but which have in common a
particular emphasis on decision-making.

ANALYSING DECISIONS IN CONFLICT

   One way of conceptualising conflict is to concentrate on
the actors involved and the actions each can take. Here,
the influence of the mathematical Theory of Games continues to
exert a strong influence, dating from J. Von Neumann and
O. Morgenstern's classic "Theory of Games and Economic
Behavior", first published in 1944. As the title suggests,
this approach was first seen primarily as a means of analysing
economic competition: however it has been widely applied
elsewhere [1], particularly to the study of international
conflict. In many ways, the terminology used is unfortunate:
to describe (for example) nuclear war as a "game" can seem
grotesque to say the least. It should therefore be stressed
that no sense of trivialisation is implied. Here, a "game"
is simply a situation having a certain structure of relevant
participants ("players"), each with various courses of action
available ("moves", making up "strategies"), and particular
preferences for the possible outcomes. The issues at stake
may be great or small: nor need one player's loss be another's
gain. In focussing on decision-makers and their possible
courses of action, this is sometimes characterised as a
"rational actor" approach, though if so, "rationality" is
sometimes defined very broadly. (That this is still only one
way of looking at conflict will become clear. from papers in
other sections.) The basic approach of Game Theory has been
developed in many directions, several of which are reviewed
and illustrated in Professor Lyn Thomas's opening paper. There
follow two papers - Professor Ken Binmore's and my own - which
start from real or alleged inadequacies in the theory and
offer two partly compatible, partly contrasting responses to
them. The remaining three papers describe ways of dealing

1
It is to be regretted that a paper presented to the conference
by Dr. A. Grafen on the application of game theory to a quite
different area, that of biological evolution, could not appear
here . However, references to work in this field may be found
in the papers by Binmore and Zeeman.

with other dimensions of conflict situations, still with an actor/decision orientation. Using the situation of a hijack as an illustration, Professor Chris Zeeman uses the relatively new mathematics of Catastrophe Theory to analyse how gradual shifts in belief and attitude may produce sudden, discontinuous changes in behaviour. Dr. Simon French considers the problem of the decision-maker faced with conflicting advice about the probability of certain events. What rules might he reasonably use to resolve the conflicting estimates he receives? Finally, the paper by Dr. Maurice Yolles shows how the underlying framework of Game Theory might be expanded in order better to deal with dynamic and stochastic aspects of conflicts.

ARMS RACES AND INTERNATIONAL SYSTEMS

The use of mathematical models to study international affairs was pioneered by Lewis Fry Richardson. His analysis of the dynamics of arms races, famously first published just prior to the outbreak of World War II, is a classic in the field. This, and Richardson's later work, is neatly summarised in Professor Ian Sutherland's opening paper, which also outlines areas for further research. The second paper, by Dr. Paul Smoker of the Richardson Institute, describes how the challenge of developing conflict analysis as a social science has been taken up, in particular through the computer simulation of complex international conflicts. In the course of this, the paper also provides a valuable introduction to many of the more philosophical issues surrounding conflict and peace research. Next, Dr. Ib Petersen describes the use of a probabilistic model to study communication and conflict within international systems, with applications to several historical cases- work relating closely to the theoretical developments outlined in the earlier paper by Yolles. To end this section, Dr. Steven Salter argues for a set of proposals designed to halt, and then reverse, the current East-West arms race. This leads us back to game analogies: the emphasis now, however, is how to avoid becoming trapped in a stupid game that may benefit no-one. His proposals use a model of the problem analogous to that of dividing up a cake between two parties unable to agree about the value of different portions.

MODELLING BATTLE

The final section, on the modelling of battle, is arguably the most self-contained. If the development of Game Theory is inescapably linked with the name of Von Neumann, and the study of arms races with that of Richardson, the use of mathematics to study battle will always be associated with F.W. Lanchester. In work originating during World War I, Lanchester pioneered mathematical models of attrition during exhange of fire. An early result for which this work is still famous is the "square law" relating the overall fighting strength of a unit to its numerical strength. This approach, and others, are summarised in the opening paper by Doug Andrews and Gordon Laing. They also discuss the general rationale underlying "Defence Operational Analysis", a theme taken up in the set of observations provided by Dr. Terence Price, a former director of the Defence Operational Analysis Establishment. Lanchester's model provides a powerful argument for concentrating one's forces in large numbers. As evidenced by Professor Robert Neild's following paper, such calculations are not only of interest to the military technician, but have wider political implications. Arguing that high-precision weapons have undermined the arguments for concentration, Neild pursues proposals for a dispersed defence which, if accepted, would revolutionise the whole military posture of the NATO countries. The next two papers are a little more technical in nature. It is noteworthy that the modelling of battle has to some extent mirrored that of arms races, despite the different orientations of the two lines of research - and the very different resources that have been made available to them. Both started with a simple, deterministic analytical model - expressed in one case by Lanchester's original equations, in the other by Richardson's. In the search for greater realism, the models have become more and more complex, until sophisticated simulation techniques are apparently needed. However, model complexity brings its own set of difficulties, and it may be preferable to obtain approximate results more simply. The paper by Peter Haysman and Kathryn Wand shows how approximate analytical solutions can be obtained for a class of models otherwise requiring simulation techniques. That by Gordon Laing describes how standard spreadsheet packages can be used to analyse, at least approximately, a range of apparently very complex problems. Lastly, the whole methodological basis of Defence Operational Analysis is subjected to a critical and lengthy scrutiny by W.T. Lord.

This book does not claim to give a complete picture of the contribution of mathematics to the study of conflict. As will be seen, some important techniques could only be mentioned in passing. Even less does it cover all fields of application. In one way or another, conflicts at an international level tend to predominate here, though much interesting work exists in other areas. Even then, specific problems such as that of nuclear deterrence, on which a voluminous and partly mathematical literature exists, are touched on only occasionally. The intention is rather to provide a introduction to current work in chosen areas, and also some indication of the variety of ideas, methods and models available.

In putting together the book, it has not been my intention
merely to reproduce the presentations given at Cambridge. All
the papers were produced after the conference, which itself
had provided much food for thought. While some approximate
fairly closely to those originally presented, the majority
have been considerably amended. An editorial aim has been
for papers to be as far as possible comprehensible to the
proverbial "intelligent layman" with no specialist knowledge
of the areas discussed. For contributors, the difficulty of
this task varied greatly according to the subject matter.
Thanks are particularly due to those who drafted and
re-drafted papers in response to repeated editorial nagging.
Their patience and effort is greatly appreciated.

Finally, I should like to take this opportunity to thank,
as well as all the contributors, Julian Hunt for his
organisation of the conference, and Catherine Richards and her
team at the IMA for all their hard work, both in setting up the
conference itself, and especially in the preparation of this
book. I hope that the result is a worthy reflection of all
their efforts.

Peter Bennett
July, 1986

ACKNOWLEDGEMENTS

The Institute thanks the authors of the papers, the editor, Dr. P.G. Bennett (University of Sussex) and also Miss Karen Jenkins, Miss Deborah Brown and Miss Pamela Irving for typing the papers.

CONTRIBUTERS

D.R. ANDREWS; *Defence Operational Analysis Establishment, Parvis Road, West Byfleet, Weybridge, Surrey, KT14 6LY.*

P.G. BENNETT; *Operational Research, Mathematics and Physics Building, University of Sussex, Falmer, Brighton, BN1 9QH.*

K.G. BINMORE; *The London School of Economics and Political Science, University of London, Houghton Street, London, WC2A 2AE.*

S. FRENCH; *Department of Mathematics, The Manchester-Sheffield School of Probability and Statistics, Statistical Laboratory, The University, Manchester, M13 9PL.*

P.J. HAYSMAN; *ROF Future Systems Group, Royal Military College of Science, Shrivenham, Swindon, Wilts., SN6 8LA.*

G.J. LAING; *Defence Operational Analysis Establishment, Parvis Road, West Byfleet, Weybridge, Surrey, KT14 6LY.*

W.T. LORD; *Rex, Thompson & Partners, Farnham, Surrey.*

R. NEILD; *Trinity College, University of Cambridge, CB3 9DU.*

IB PETERSEN; *Institute of Political Studies, University of Copenhagen, Rosenborggade 15, DK-1130, Copenhagen, Denmark.*

T. PRICE; *The Uranium Institute, Bowater House, 68 Knightsbridge, London, SW1X 7LT.*

S.H. SALTER; *Department of Mechanical Engineering, University of Edinburgh, King's Building, Edinburgh, EH9 3JL.*

P. SMOKER; *The Richardson Institute for Conflict and Peace Research, Department of Politics, University of Lancaster, Lancaster, LA1 4YF.*

I. SUTHERLAND; *MRC Biostatistics Unit, Medical Research Council Centre, Hills Road, Cambridge, CB2 2QH.*

L.C. THOMAS; *Department of Business Studies, University of Edinburgh, George Square, Edinburgh.*

K. WAND; *Operational Research and Statistics Group, Royal Military College of Science, Shrivenham, Swindon, Wilts., SN6 8LA.*

M. YOLLES; *The Richardson Institute for Conflict and Peace Research, Department of Politics, University of Lancaster, Lancaster, LA1 4YF.*

E.C. ZEEMAN; *Department of Mathematics, University of Warwick, Coventry, CV4 7AL.*

1. ANALYSING DECISIONS IN CONFLICT

Moriarty: "... you see, the United Anti-Socialist Neo-
            democratic Pro-Fascist Communist Party are fighting
            to overthrow the Utilateral Democratic United
            Partisan Bellicose Pacifist Cobelligerent Tory
            Labour Liberal Party."

Seagoon: "Whose side are you on?"

Moriarty: "There are no sides - we're all in this together..."

(The Goon Show Scripts, Spike Milligan, Woburn Press 1972)

USING GAME THEORY AND ITS EXTENSIONS TO MODEL CONFLICT

L.C. Thomas
(Department of Business Studies, University of Edinburgh)

1. INTRODUCTION

Game theory has its roots in the analysis of conflict as well as in economic modelling, and its extensions also look to these areas for applications. We examine two such extensions, metagame theory and multi-stage games and look at how some very simple examples of conflict can be modelled using these techniques. Our aim is to show that metagame analysis and the analysis of supergames, which are a special type of multi-stage game, are closely related, and that using multistage games enables one to illuminate some of the problems in conflict that ordinary game theory finds difficult to deal with.

In section two we give a very brief sketch of the history of game theory to put its extensions into context. Section three deals with metagame analysis by applying it to the Cuban missile crisis and the Falklands war. In section four we introduce supergames and show that the outcomes of a basic game which are supergame stable are related to these which are stable in the metagame analysis. Lastly in section five, we discuss, by looking at examples, three areas where multi-stage games might prove useful by concentrating on important features in the conflict. All three are related to the point that conflicts take time to unfold and this time factor and the changes that occur because of it are important.

2. GAME THEORY AND ITS EXTENSIONS

It is usual to date the start of game theory with the publication of von Neumann and Morgenstern's Theory of Games and Economic Behaviour" [32]. This book had an euphoric reception, and there was a feeling that game theory was going to solve many of the problems of conflict analysis.

The terminology set down in this book has been standard for
the subject. Game theory consists of ways of analysing models
of conflict of interest between two or more parties. The
participants in the game are termed players and each game
consists of a sequence of moves which are either decisions by
the players or outcomes of chance events. A strategy for a
player is a list of the decisions he will make at all possible
situations that can arise in the game. Think of it as a set
of instructions which enables a computer to play the game for
you. Obviously in many games - chess for example - although
in principle you can conceive of a strategy, in practice it is
too long to write down. Whatever strategy each player chooses
at the end of the game each one receives a payoff. This could
be a numerical payoff representing his actual award, or the
utility of the outcome of the game to him, or an indication of
the preference ordering of this outcome amongst all possible
outcomes of the game for that player. If the sum of the
players' outcomes is zero no matter what strategies they use, the
game is called zero-sum. These are games like Poker, Chess,
etc. where what one player wins, the other loses. Games which
do not have this property are called, not surprisingly, non-
zero-sum.

During the early 50's there was a great deal of work in the
game theory area and the four volumes of Contributions to the
Theory of Games, Kuhn and Tucker [17,18], Drescher, Tucker
Wolfe [6], Tucker and Luce [31] give an account of the variety
of problems looked at. However, as the difficulties with
modelling non-zero-sum games and those with more than two
players become apparent, there was a realisation of the
limitations of game theory. This is obvious in the other
classic text-book of game theory by Luce and Raiffa [19]
written in 1957. The idea of playing a game over and over
again to examine the stability of the strategies was
introduced there, but was then taken up by Aumann [1, 2] and
Aumann and Maschler [3] and used to investigate possible
solution concepts for n-person games and also games where the
players only have partial information. The latter is still
an area of active interest see Kohlberg [16] while Friedman
[9] used the ideas of supergames to deal with price setting in
oligopolistic competition.

The theory of metagames expounded by Howard [11, 12, 13]
also has its roots in von Neumann and Morgenstern's book but
also takes something from the work of Rapoport [23], Boulding
[5] and Schelling [25] who use several different models and
techniques, including game theory and the work of Richardson,
to analyse conflict problems. Howard [14] transformed
metagames to analysis of options by giving a recipe for
checking whether an outcome was a metaequilibrium, while Fraser

and Hipel [8] also wrote a computer algorithm to find the
metaequilibria in real conflict problems. In metagames, one
is interested in trying to guess an opponent's strategy or
guess his guess of your strategy. This idea was expanded by
Bennett [4] to allow for the possibility that the players
might have completely different perceptions of the game even
down to differing about the number of strategies a player
might have. The analysis of these hypergames is closely
related to the analysis of metagames, so much so that
Takahashi, Fraser and Hipel [24] use their metagame algorithm
to analyse a hypergame.

A brief introduction to metagames and supergames, is found
in Thomas [30], and a bibliography of applications of game
theory in conflict is given by Intriligator [15].

3. EXAMPLES OF METAGAMES USED IN CONFLICT ANALYSIS

One of the ideas that led Howard [11, 12, 13] to introduce
metagame theory was the assumption that each player tries to
predict which strategy his opponent will choose and then
chooses the outcome which is most favourable to him. A stable
outcome is one where all the players correctly predict the
other players' strategies. We must be careful of our
terminology here though, because a strategy for a player, could
be of the form: if my opponent does A, I will play X; if he
plays B, I will play Y, etc.

Howard identified these stable outcomes as the equilibria of
metagames - enlarged games based on the original one G. 1G is
the game where player 2 announces his action in G first and 1
chooses his in the knowledge of this choice. Such games were
first discussed in von Neumann and Morgenstern [32], who
called them majorant and minorant games. 21G is the metagame
where first 1 announces his strategy in the game 1G. This
strategy is of the form 'if 2 does X in G, I(1) will do A; if
2 does Y, I will do B". 2 then chooses his action in G, in
the knowledge of which reaction function 1 will choose.
Similar definitions hold for 2G and 12G, and these games are
supposed to represent the cases where 2 knows (i.e. correctly
predicts) 1's action in G and 1 knows or correctly predicts
2's reaction function.

Howard [13] showed that if G is a 2-person game it is
enough to look at 12G and 21G to find all the stable outcomes
(metaequilibria). If $e_i(x,y)$ is the payoff to player i when 1
plays strategy x in G and 2 chooses strategy y, he also showed
that $(\bar{x},\bar{y})$ is a metaequilibrium arising from 12G if

$$e_1(\bar{x},\bar{y}) \geq \max_{x} \min_{y} e_1(x,y)$$

$$e_2(\bar{x},\bar{y}) \geq \min_{x} \max_{y} e_2(x,y)$$

(3.1)

and is a metaequilibrium arising from 21G if

$$e_1(\bar{x},\bar{y}) \geq \min_{y} \max_{x} e_1(x,y)$$

$$e_2(\bar{x},\bar{y}) \geq \max_{y} \min_{x} e_2(x,y)$$

(3.2)

The symmetric metaequilibria which satisfy both (3.1) and (3.2) correspond to outcomes which are stable, no matter which way we think of the game. Such $(\bar{x},\bar{y})$ must satisfy

$$e_1(\bar{x},\bar{y}) \geq \min_{y} \max_{x} e_1(x,y)$$

$$e_2(\bar{x},\bar{y}) \geq \min_{x} \max_{y} e_2(x,y)$$

(3.3)

As an example of a metagame analysis, consider the Cuban Missile crisis in 1962. The U.S. government had discovered that the U.S.S.R. had missiles stationed in Cuba and made preparations for an invasion of Cuba to remove the missiles. The U.S. had two courses of action - to abandon its invasion or to continue it. The U.S.S.R. could withdraw its missiles or maintain them in Cuba. If we order the outcomes 4,3,2,1 in order of decreasing preference, a realistic ordering might be

U.S.S.R. missiles

		withdraw (w)	maintain (m)
U.S.	abandon (a)	(3,3)	(2,4)
invasion plans	continue (c)	(4,2)	(1,1)

The equilibrium in the basic game G, $E(G) = \{(a,m)\ (c,w)\}$, which is a 'victory' for the U.S.S.R. or the U.S. However, using (3.1) and (3.2) it follows that $\hat{E}(12G)$ and $\hat{E}(21G)$ the sets of metaequilibria arising from 12G and 21G respectively are

$$\hat{E}(12G) = \hat{E}(21G) = \{(a,m),\ (c,w),\ (a,w)\}$$

The extra outcome - (abandon, withdraw) -, which was what
actually occurred arises from the strategies in which the U.S.
says "if you withdraw, I will abandon; otherwise I will
continue", and the U.S.S.R. "maintains the missiles unless the
U.S. abandons its invasion".

This is the same sort of game as the game of Chicken -
popular with teenagers in the 1950's where two cars drove at
each other - and the one to swerve was 'chicken' and 'lost'.
The preference, 'I don't swerve, you swerve, is better than
we both swerve is better than, I swerve you do not, is better
than we both do not swerve, 'leads to the same problem.

The negotiations at the outbreak of the Falklands War,
appeared on the surface to have very much in common with the
Cuban Missile crisis and yet the outcome was completely
different. In 1982, Argentina successfully invaded the
Falkland Islands, and deported the British administration.
The U.K. government then prepared a Task Force to retake the
islands. At this point the options open to each side were the
same as in the missile crisis - the United Kingdom could
continue or abandon its invasion; the Argentine forces could
either withdraw or stay on the island. A possible preference
structure might be

Argentine forces

		withdraw (w)	maintain (m)
U.K.	abandon (a)	(4,2)	(1,4)
invasion	continue (c)	(3,1)	(2,3)

It is not to be assumed that this was the actual preference
structure, but just one reasonable interpretation, which is
used for illustration.

The only equilibrium pair in this game is (c,m) and
substituting the facts that $\max_x \min_y e_1(x,y) = \min_y \max_x e_1(x,y)$
$= 2$, and $\min_x \max_y e_2(x,y) = \max_y \min_x e_2(x,y) = 3$ into (3.1) and
(3.2) gives us that only the metaequilibria in 12G or 21G is
(c,m). One might argue with the preference orders displayed,
but in fact if you say that the U.K. would prefer (c,w) to
(a,w) then the analysis remains unchanged. Metagame theory
does not allow any other stable outcome than what actually
occurred.

A closer examination of the game shows in fact that the
Argentine preference to maintaining their forces on the island
means there cannot be any stable outcome which does not
involve this choice by them. In order to avoid a war it is
necessary to persuade the Argentine government to reverse
their preferences between 'continue-maintain' and 'abandon-
withdraw'. If this could be done the preferences lead to:

Argentine forces

		c	w	m
U.K.	a	$(4,3)$		$(1,4)$
	c	$(3,1)$		$(2,2)$

Although (c,m) is still the only equilibrium in the basic
game, the metagames lead to metaequilibria.

$E(21G) = E(12G) = \{(a,w), (c,m)\}$. The abandon-withdraw
outcome is obtained by threatening to continue the invasion
if the forces were maintained on the island but to abandon if
they withdraw and vice versa for the Argentine forces.

We will return to this example later, but first it is
necessary to point out the close connection between metagame
analysis and that of a supergame formulation of the problem.

4. SUPERGAMES AND METAGAMES

Luce and Raiffa [19] introduced the term supergame, to
describe the repeated playing of the same game. The object
of such an analysis was to find which pairs of strategies
were likely to be played over and over again. By looking at
what payoffs the equilibrium pairs of strategies in the
supergame actually give rise to in the basic game, one could
get a feel for the stability of the various outcomes. This is
obviously very closely related to the aims underlying metagame
theory, so it is not surprising that the results are closely
related.

Since a supergame G^s involves an infinite number of plays
of the basic game G, the total payoff to a player could also
well be infinity. We want to avoid this as we cannot compare
two strategies which give rise to infinite payoffs, so instead
of the total payoffs, we take the more discriminating criteria
of average payoff per game or total discounted payoff.

A typical strategy for player 1 in a supergame is

$$X = \{x(1), \; x(2,H_1), \; x(3,H_2), \; \ldots \; x(n,H_{n-1}) \; \ldots\}$$

where $x(n_1, H_{n-1})$ is the strategy played by player 1 at the nth playing of the game G, and H_{n-1} describes what has occurred in the first n-1 stages. If Y is a similar strategy for player 2, and $e_i(.,.)$ is the payoff matrix for player i in the basic game G, the total discounted payoff in G^s for player i is defined by

$$E_i^\beta(X,Y) = \sum_{k=1}^{\infty} \beta^{k-1} e_i(x(k,H_{k-1}), y(k,H_{k-1})) \qquad (4.1)$$

where β is the discount factor.

The average payoff per game is defined as

$$E(X,Y) = \lim_{n\to\infty} \left(\sum_{k=1}^{n} e_i(x(k,H_{k-1}), y(k,H_{k-1})) \right)/n \qquad (4.2)$$

(If the limit does not exist we can use limsup or liminf with no change in subsequent results.)

We are interested in the equilibrium pairs of the supergame and in particular the pairs that result in the same pair of strategies being played at every playing of the basic game G. A pair of strategies (x,y) in the basic game G is called supergame stable if there is an equilibrium pair (X,Y) in the supergame G^s which result in (x,y) being played at every stage of the supergame.

We are now in a position to show how closely the ideas of stability in supergames and metagames are related.

Theorem 4.1. If G is a two-person non-zero-sum game with finite number of pure strategies for each player, then the supergame stable strategies for the average payoff per stage supergame G^s are equal to the symmetric metaequilibrium of the metagames based on G.

Details of the proof are given in the appendix, but the important point is to notice that a pair (x,y) which satisfies (3.3) is supergame stable because it results from the equilibrium pair (X,Y) in G^s being played where if $X = \{x(n, H_{n-1}), n = 1, 2, ...,\}$ and $Y = \{y(n, H_{n-1}), n = 1, 2, ...\}$

$$x(n, H_{n-1}) = x \text{ if } H_{n-1} = \{(x,y)\}^{n-1}, \text{ i.e. } (x,y) \text{ is played at all}$$

previous n-1 stages

$$= \bar{x} \text{ otherwise where } \min_x \max_y e_2(x,y) = \max_y e_2(\bar{x},y)$$

(4.3)

and

$$y(n, H_{n-1}) = y \text{ if } H_{n-1} = \{x,y)\}^{n-1}$$

(4.4)

$$= \bar{y} \text{ otherwise, where } \max_x e_1(x,\bar{y}) = \min_y \max_x e_1(x,y)$$

Thus if either player deviates from the pair (x,y), the other player will 'punish' him for evermore by playing the minimax strategy that limits the first player's payoffs as much as possible.

A similar though not identical result holds for the discounted version of the supergame.

Theorem 4.2. If G is a two-person non-zero sum game with finite number of pure strategies for each player, then for any pair of strategies $(\bar{x},\bar{y})$ in G such that

$$e_1(\bar{x},\bar{y}) > \min_y \max_x e_1(x,y); \ e_2(\bar{x},\bar{y}) > \min_x \max_y e_2(x,y)$$

(4.5)

there is a $\beta^* < 1$, so that for all $\beta \geq \beta^*$ (x,y) is supergame stable in the discounted supergame with discount factor β.

The proof follows very closely that of Theorem 4.1 (see appendix). The result says that the supergame stable strategies for discounted supergames are essentially the interior of the set of metagame equilibria.

5.   MULTI-STAGE GAMES USED TO MODEL CONFLICT

In the previous section, we showed how the results of a metagame analysis and a supergame analysis of conflict are closely related. However supergames are only a very special case of multi-stage games. In a supergame, one plays the same game over and over again, while in multistage games one is allowed to play different games at each stage. The choice of the game to play at the next stage can depend on the game played at this stage, the strategies played in it and the outcome of a random event. The two player zero-sum discounted version of such multistage games (also called stochastic or

Markov games) was introduced in a classic paper by Shapley [26] who showed they always had a value and optimal strategies and described an algorithm for solving them. The average reward version involves more subtle problems and it was only recently that Mertens and Neyman [20] and Monash [21] independently proved that all such games have a value provided there are only a finite number of different subgames that can be played and each subgame has only a finite number of pure strategies. Even so, there may not be optimal strategies which guarantee the value. Rogers [24] proved that in the non-zero sum case, there always exists at least one equilibrium pair of strategies.

There are many difficulties in using game theory to model conflict problems, but multi-stage games do enable one to address some of them. Here we will indicate three types of difficulties which may be illuminated when modelled in multi-stage games.

A.   Changes in preferences during the course of the game.

One of the criticisms levelled against game models of situation is that it is necessary to ascribe values to all the outcomes in advance and so define a preference relationship for each player over the outcomes of the game. This may be difficult to do when it is not obvious initially what is exactly involved in some of the outcomes. Also it is quite possible that events that occur during the game will change players' preferences between the various outcomes. Ideally we might wish to develop a form of analysis, where a player's preference between courses of action did not have to be defined until it was necessary to perform these actions, but obviously it would then be impossible to perform such an analysis beforehand, or obtain any idea of a good overall strategy. If one has to state the player's preference beforehand then modelling the problem as a multi-stage game concentrates one's mind on the effects that events in the game might have on subsequent preferences between actions.

As a very simple example of this use of multi-stage games, let us return to our model of the Falklands War in section three. We will assume there are two subgames, G_1 before there have been any major hostilities and G_2 after hostilities and consequent loss of life have occurred. The outcomes of G_1 are either settlements of the conflict before hostilities or moving on to playing the game G_2. We will assume that the payoffs representing the preferences are

G_1 Argentine forces

                                  withdraw w   maintain m

U.K.   abandon (a) $\begin{pmatrix} (7,5) & (3,7) \\ (6,2) & G_2 \end{pmatrix}$

invasion   continue (c)

                                                              (5.1)

G_2 Argentine forces

                                  withdraw w   maintain m

U.K.   abandon (a) $\begin{pmatrix} (2,4) & (1,6) \\ (5,1) & (4,3) \end{pmatrix}$

invasion   continue (c)

                                                              (5.2)

It is not my intention to defend the preferences given above, although the idea that once hostilities had commenced the U.K. government wanted the invasion to continue seems a tenable proposition. Nicholson [22] has some pertinent points to make on the factors that are necessary in order for a country to become irrevocably committed to continue a war. The ordering 7 to 1 is to represent the preferences in decreasing order of the players within and between the games. In practice, it would seem sensible to order the payoffs within each game first and then combine the two orders. Notice that if we were to replace G_2 by the payoff (4,6) - which are

possible payoffs for each player in G_2, we would have the same

ordering as we took when we introduced the example in section three.

We can analyse this by examining G_2 first and substituting the 'solution' of G_2 as the payoff of (c,m) in G_1 or

alternatively write out the strategies for the overall game, i.e. "abandon in G_1" (a) "continue in G_1 but abandon in G_2"

(c,a), and "continue in G_1 and G_2", (c,c) and the resultant

payoff matrix, which is

$\overline{G}$ (w)   (m,w)   (m,m)

U.K.   (a) $\begin{pmatrix} (7.5) & (3,7) & (3,7) \\ (6,2) & (2,4) & (1,6) \\ (6,2) & (5,1) & (4,3) \end{pmatrix}$

          (c;a)   (5.3)

          (c;c)

In this case both methods lead to the same result. The only equilibrium pair in G_2 is (c,m) and substituting its payoff (4,3)) in G_1 gives (c,m) as the only payoff, while ((c;c), (m;m)) is the only equilibrium pair in $\bar{G}$. If we find the metaequilibria of the respective games, we find that $E(12G_2) = E(21G_2) = \{(c,m)\}$ and substituting its payoff (4,3) in G_1 leads to $E(12G_1) = E(21G_1) = \{(a,w); ((c;c), (m;m))\}$. Similarly $E(12\bar{G}) = E(21\bar{G}) = \{(a;w), ((c;c), (m;m))\}$.

It is interesting to note that this analysis suggests that abandon-withdraw is a stable strategy. It corresponds to the negotiation positions where the U.K. says, "We will abandon if you withdraw, otherwise we will continue with the invasion even after hostilities have commenced", while the Argentine says, "we will withdraw if you abandon the invasion; otherwise we will keep our forces there no matter what happens". This outcome is stable now because both sides' preferences have changed after hostilities have begun and in particular the Argentines prefer to (abandon-withdraw) before hostilities begin to (continue-maintain) after hostilities have occurred.

b) Timing of Actions

Another facet of conflict that is more easily modelled by multi-stage games, than by the normal form of standard games, is the timing of action. It is sometimes as important to decide when an action is taken as which action is taken. On other occasions, the passage of time, even though no action is taken by either player confers an advantage on one or other of them. Naively, one could view a strike in this way. It is a method of delaying or even avoiding negotiations, in such a way that confers an advantage on one of the players. Again we will take an over-simplified example to describe the use of multi-stage games.

Suppose a union would either accept a low or high wage rise, and the management also can be prepared to accept either a low or high rise. The payoffs to the two sides, (not just their preference orders this time) are

$$
\begin{array}{cc}
 & \text{Union} \\
 & \begin{array}{cc} \text{Low} & \text{High} \end{array}
\end{array}
$$

		Low	High	
	High	(2,2)	(1,3)	
Management				(5.4)
	Low	(3,1)		

One could argue that if the union were submitting a low wage
claim and the management were willing to accept a high one, the
payoff should be (3,1) since the low claim will be accepted.
The defence of a payoff (2,2) is that both sides would ask
for more (or less) than they would accept, and in this case, a
compromise solution is likely. However the subsequent analysis
is not affected if we replace (2,2) by (3,1). Assume this is
the game G_1 in the first period and that if management will

only accept a low rise and the union a high one, there will be
a strike and the game goes to the second period G_2. Similarly
we define the payoffs in the nth period G_n as

$$
G_n \qquad\qquad\qquad \text{Union}
$$

$$
\begin{array}{c}
 \\
\text{Management} \\
\end{array}
\begin{array}{c}
 \\
\text{High} \\
\text{Low}
\end{array}
\begin{array}{cc}
\text{Low} & \text{High} \\
\left(\begin{array}{cc}
(2,2) & (1,3) \\
(3,1) & G_{n+1}
\end{array} \right)
\end{array}
\qquad (5.5)
$$

Suppose each period there is no settlement costs each side a
diminution of 1 in payoff. This displays again the
difficulties of ascribing realistic values to the payoffs
because we are subtracting a short-term loss from the long-
term effect of the wage settlement. However, for the
purposes of simplicity we will ignore such difficulties here.

G_n, then has payoffs

$$
\begin{array}{c}
\text{High} \\
\text{Low}
\end{array}
\begin{array}{cc}
\text{Low} & \text{High} \\
\left(\begin{array}{cc}
(2-n,2-n) & (1-n,3-n) \\
(3-n,1-n) & G_{n+1}
\end{array} \right)
\end{array}
\qquad (5.6)
$$

We can write the strategies for the overall game for the
Management as H, LH, LLH, etc. where LLH means only accept a
low pay rise in the first two periods and a high one
subsequently. Similar strategies occur for the union and the
payoff matrix becomes

Union

		L	HL	HHL	HHHL	HHHH
	H	(2,2)	(1,3)	(1,3)*	(1,3)*	(1,3)*
	LH	(3,1)	(1,1)	(0,2)	(0,2)	(0,2)
Management	LLH	(3,1)*	(2,0)	(0,0)	(-1,1)	(-1,1)
	LLLH	(3,1)*	(2,0)	(1,-1)	(-1,-1)	(-2,0)
	LLLL	(3,1)*	(2,0)	(1,-1)	(0,-2)	G_5

(5.7)

The equilibrium pairs (starred in 5.7) are of two forms – one where management must play low for at least two periods then high at some point against a union that plays low, and secondly where management settles high against a union that will demand high for at least two periods. The idea is that the belief that the other side will be prepared to strike for at least two weeks, justifies a capitulation now. However, since the game is symmetric, we are left with the dilemma of which of two very different sets of equilibrium pairs is likely to occur.

If the cost of a strike is not equally disadvantageous to both sides – say it costs 3 for the management and 1 for the union, the symmetry is broken and we are led to a payoff matrix:

		L	HL	HHL	HHHL	HHHH-
	H	(2,2)	(1,3)*	(1,3)*	(1,3)*	(1,3)*
	LH	(3,1)	(-1,1)	(-2,2)	(-2,2)	(-2,2)
Management	LLH	(3,1)*	(0,0)	(-4,0)	(-5,1)	(-5,1)
	LLLH	(3.1)*	(0,0)	(-3,-1)	(-7,-1)	(-8,0)
	LLLL-	(3,1)*	(0,0)	(-3,-1)	(-6,-2)	G_5

Here the equilibrium pairs consist of Management playing high and Union demanding a high pay claim for at least one period or the union settling for a low claim if the management is prepared to offer low for at least two periods. The fact that the union has only to persuade the management it will strike for one period, to get to an equilibrium situation, but the management needs to persuade the union it is prepared to endure at least a two period strike to get to its preferable equilibrium pair, puts the union at an advantage.

One difficulty with this analysis is that it takes no account of 'sunk costs' that the players have already invested. The same analysis would be made whether it was the first period of the strike or the twenty-first period of the strike in which it was made. One way of allowing for this, is to put a limit on the loss each side can take because of the strike before they have to settle. Suppose in the last example, both unions and management can only take a loss of 6, then the management can only afford a two-period strike before they have to settle. Thus we have the following multi-stage game

$$G_1 \quad\quad \begin{array}{cc} L & H \end{array}$$
$$\begin{array}{c} H \\ L \end{array} \begin{pmatrix} (2,2) & (1,3) \\ (3,1) & G_2 \end{pmatrix}$$

$$G_2 \quad\quad \begin{array}{cc} L & H \end{array}$$
$$\begin{array}{c} H \\ L \end{array} \begin{pmatrix} (-1,1) & (-2,2) \\ (0,0) & G_3 \end{pmatrix}$$

$$G_3 \quad\quad \begin{array}{cc} L & H \end{array}$$
$$\begin{array}{c} H \\ L \end{array} \begin{pmatrix} (-4,0) & (-5.1) \\ (-3,-1) & G_o \end{pmatrix}$$

$$G_o \quad\quad \begin{array}{cc} L & H \end{array}$$
$$\begin{array}{c} H \end{array} \begin{pmatrix} (-4,0) & (-5,1) \end{pmatrix}$$

where G_o is where the Management has to settle immediately and so agrees to a high wage rise. The solution of G_o must be (H,H) for a payoff $(-5,1)$ and substituting this into G_3, shows that the only equilibrium pair in G_3 is (L,H) with value $(-5,1)$. Repeating this procedure, the only equilibrium pair on G_2 is (H,H) with payoff $(-2,2)$, and then the only one in G_1 is also (H,H) with payoff $(1,3)$. Obviously an analysis which depends on finding the equilibrium pairs in a non-zero sum game is open to a great deal of criticism, see Luce and Raiffa [19] for a classical exposition of the problem. However using a multi-stage game approach does highlight how time can affect the outcome of a game.

c) Differences in analysis made a priori and during the game

The third area where multi-stage game formulation is useful is in clarifying the difficulties in games where one is tempted to take an action for the short-term advantage it involves without appreciating the long-term disadvantages. This tends to occur when one finds oneself in the middle of a conflict situation without realising it and not having been able to make

an a priori analysis of the conflict. So one might be in a
situation, which would not have arisen if one could have
analysed the problem originally. Alternatively, the 'sunk
costs' that one has already invested in the conflict would lead
one to a different decision than would be made a priori.

The dollar auction of Shubik [27] is a classical example of
such a case and is also taken as a model of certain forms of
arms escalation. (Salter's paper [28] in this volume is
another example.) In the dollar auction, two players are
invited to bid for a dollar in bids which are multiples of 5
cents and the one who bids the more wins it. The catch is that
both have to pay the highest amount they bid. This is the same
problem as the 'war of attrition' models in evolutionary games
(see Haigh and Rose [10]).

Suppose player 1 has bid i cents and player 2 has just bid
j(j > i)cents. What should player 1 do? If he does not bid,
he will lose i cents; if he bids k cents, k>j, he stands to
win 100 - k cents, which provided k< i + 100 is better than
-i cents. So he will bid j+5 cents, as this promises to give
him the best profit. Player 2 will then perform the same
analysis and bid another 5 cents irrespective of the bid
already made. Thus the bids will not stop at one dollar but
apparently continue on upwards.

Representing it as a multi-stage game, let $G_{i,j}^{1}$ be the game
when 1 has bid i and 2 bid j > i then the payoff matrix is

$$\begin{array}{l} \text{no bid} \\ \\ \text{bid}_{j+5k} \end{array} \begin{pmatrix} (-i, & 100-j) \\ \\ G_{j+5k,j}^{2} \end{pmatrix} \text{ for all k = 1, 2, etc.}$$

G_{ij}^{2} defined similarly but with i > j has payoff matrices

$$\begin{array}{cc} \text{No bid} & \text{bid} \\ ((100-i, -j), & G_{i, i+5k}^{1}) \end{array} \quad k = 1, 2, \ldots,$$

This is then a recursive game (see Everett [7]) and is an
example of a game that does not end. If you are able to
analyse the game a priori, the best suggestion is for player 1
to bid exactly one dollar, and for player 2 not to bid at all.
This gives both players zero payoff. If one actually finds
oneself involved in such a game, the most realistic strategy
would seem to be to "cut one's losses" and to stop, otherwise

the game will continue forever.

Summing up, it has been obvious for many years that game theory is not the panacea for solving all the problems of conflict analysis. However classical game theory and its extensions does seem to be one of the most versatile tools in modelling such problems and in enabling one to focus on the vital issues in the conflict.

APPENDIX

Proof of Theorem 4.1

Take X,Y to be the strategies in G^S defined in (4.3) and (4.4). Then $E_i(X,Y) = \lim_{n \to \infty} (\sum_{k=1}^{n} e_i(x,y))/n$, so it is enough to show (X,Y) is an equilibrium pair. Take any other strategy X' for player 1. This will result in (x,y) being played for the first k stages (where k can range from 0 to infinity), (x',y) at $k+1^{th}$ stage and $(x'_r, \bar{y})$ r = k + 2, ... at the subsequent stages. Thus

$E_1(X',Y) - E_1(X,Y)$

$\qquad \qquad \qquad \qquad \qquad \qquad \qquad \qquad \qquad \qquad$ (A.1)

$\qquad = \lim_{n \to \infty} (e_1(x',y) - e_1(x,y) + \sum_{r=k+2}^{n} e_1(x'_r, \bar{y}) - e_1(x,y))/n$

Since $e_1(x,y) \geq \min_{v} \max_{u} e_1(u,v) = \max_{u} e_1(u,\bar{y}) \geq e_1(x'_r, \bar{y})$, thus

$E_1(X',Y) - E_1(X,Y) \leq \lim_{n \to \infty} (e_1(x',y) - e_1(x,y))/n = 0.$ So

$E_1(X,Y) \geq E_1(X',Y)$, and a similar calculation shows that

$E_2(X,Y) \geq E_2(X,Y')$. Hence (X,Y) is an equilibrium pair and (x,y) is supergame stable.

We must also prove that for any (x,y) which do not satisfy (3.3), we cannot construct an equilibrium pair (X,Y) which results in (x,y) being played all the time. Assume without loss of generality that $e_1(x,y) < \min_{v} \max_{u} e_1(u,v)$ then

in order that (x,y) are played all the time it is necessary that if $X = \{x(1), x(2,H_1), \ldots, \}$ and $Y = \{y(1), Y(2,H_1) \ldots,\}$ then

$$x(n,H_{n-1}) \quad = \quad x \quad \text{if } H_{n-1} = \{(x,y)\}^{n-1} \text{ i.e. } (x,y) \text{ for all previous games}$$

$$x_n(H_{n-1}) \text{ otherwise}$$

$$y(n, H_{n-1}) \quad = \quad y \quad \text{if } H_{n-1} = \{(x,y)\}^{n-1}$$

$$y_n(H_{n-1}) \text{ otherwise}$$

where the $x_n(H_{n-1})$, $y_n(H_{n-1})$ can be quite general. For any such Y define $X' = \{x'(1), x'(2,H_1), \ldots,\}$ by

$$x'(n,H_{n-1}) = x'_n \quad \text{where if } y'_n = y_n(H_{n-1}) \text{ for } H_{n-1} \text{ the}$$

actual history of this realisation of the game, then $e_1(x'_n,y'_n)$

$= \max\limits_{x} e_1(x,y'_n)$. Notice $e_1(x'_n, y'_n) = \max\limits_{x} e_1(x,y'_n) \geq \min\limits_{n} \max\limits_{y} \max\limits_{x}$

$e_1(x,y) > e_1(x,y)$. Thus $E_1(X',Y) - E_1(X,Y) = \lim\limits_{n\to\infty} (\sum\limits_{k=1}^{n} e_1(x'_k, y'_k)$

$-e_1(x,y))/n > \varepsilon$,

where $\varepsilon = \min\limits_{y} \max\limits_{x} e_1(x,y) - e_1(x,y)$, so (X,Y) is not in

equilibrium.

Proof of Theorem 4.2

Assume (x,y) satisfies (4.5) and define X,Y as in (4.3), (4.4)

then $E_i(X,Y) = \sum\limits_{k=1}^{\infty} \beta^{k-1} e_i(x,y)$. Again taking any other X'

against Y leads to (x,y) being played for k stages, (x',y) for the $k+1^{th}$ stage and $(x_r,\bar{y})$ at subsequent stages r. Thus

$$E_1^\beta(X,Y) - E_1^\beta(X',Y) = \beta^k(e_1(x,y) - e_1(x',y)) + \sum\limits_{r=k+2}^{\infty} \beta^{r-1}(e_1(x,y)$$

$$-e_1(x_r,\bar{y}))$$

Let $\varepsilon = e_1(x,y) - \min\limits_{y} \max\limits_{x} e_1(x,y) = e_1(x,y) - \max\limits_{x} e_1(x,\bar{y}) \geq$

$e_1(x,y) - e_1(x_r,\bar{y})$.

So $E^{\beta}(X,Y) - E_1^{\beta}(X',Y) \geq \beta^k(e_1(x,y) - e_1(x',y) + \varepsilon/(1-\beta))$

For β sufficiently close to 1, the R.H.S. of this must be positive which shows $E^{\beta}(X,Y) \geq E_1^{\beta}(X',Y)$. A similar result shows $E_2^{\beta}(X,Y) \geq E_2^{\beta}(X,Y')$ and hence (X,Y) is an equilibrium pair.

In a similar way as in Theorem 4.1 we can show that if $e_1(x,y) \leq \min\limits_{y} \max\limits_{x} e_1(x,y)$, then we cannot construct an

equilibrium pair (X,Y) which leads to (x,y) always being played.

REFERENCES

1. Aumann, R.J., (1959) Acceptable points in general
    co-operative n-person games in Contributions to the Theory
    of Games Vol. IV ed. Tucker, A.W., Luce, R.D. Princeton
    University Press pp 287-324.

2. Aumann, R.J., (1960) Acceptable points in games of perfect
    information. *Pacific J. of Maths.* **10**, 381-417.

3. Aumann, R.J. and Maschler, M., (1967) Repeated games with
    incomplete information. Report to the U.S. Arms Control
    and Disarmament Agency, Mathematics Policy Research Inc.
    pp 25-108, 287-403.

4. Bennett, P.G., (1977) Towards a theory of hypergames.
    Omega **5** 749-751.

5. Boulding, K.E., (1962) Conflict and Defense. A General
    Theory, Harper and Row, New York.

6. Drescher, M., Tucker, A.W. and Wolfe, P., (1957)
    Contributions to the Theory of Games Vol. III. Princeton
    University Press, Princeton.

7. Everett, H., (1957) 'Recursive games' in Contributions to
    the Theory of Games Vol. III, ed., Drescher, M., Tucker, A.W.
    Wolfe, P., Princeton University Press, Princeton. pp 47-48.

8. Fraser, N.M. and Hipel, K.W., (1980) 'Metagame analysis of the Poplar River Conflict'. *J. Oper. Res. Soc.* **31**, 377-385.

9. Friedman, J.W., (1977) Oligopoly and the Theory of Games, North-Holland, Amsterdam.

10. Haigh, J. and Rose, M.C., (1980) "Evolutionary game auctions", *J. Theor. Biol.* **85**, 381-397.

11. Howard, N., (1966): Theory of metagames, General Systems 11, pp 167-186.

12. Howard, N., (1966): The mathematics of metagames. General Systems 11, pp 187-200.

13. Howard, N., (1971) Paradoxes of Rationality, M.I.T. Press, Cambridge, Massachusetts.

14. Howard, N., (1973) A Computer System for foreign-policy decision making. *J. Peace Science* **1**, 61-68.

15. Intriligator, M.D., (1982): Research on conflict theory *J. Conflict Resolution* **26**, pp 307-327.

16. Kohlberg, E., (1975): Optimal strategies in repeated games with incomplete information, *Int. J. Game Theory* **4** pp 7-24.

17. Kuhn, H.W. and Tucker, A.W., (1950): Contributions to the Theory of Games Vol I, Princeton Unviersity Press, Princeton.

18. Kuhn, H.W. and Tucker, A.W., (1953): Contributions to the Theory of Games, Vol. II. Princeton University Press, Princeton.

19. Luce, R.D. and Raiffa, H., (1957) Games and Decisions, Wiley, New York.

20. Mertens, J.F. and Neyman, A., (1981) Stochastic games. *Int. J. Game Theory,* **10**, 53-66.

21. Monash, C.A., (1979): Stochastic games. The minmax theorem. Thesis, Harvard University, Cambridge.

22. Nicholson, M., (1984): Aborted wars and "prosaic" utility functions preprint, University of Manchester.

23. Rapoport, A., (1960): Fights, Games and Debates, University of Michigan Press, Ann Arbor.

24. Rogers, P.D., (1969) Non-zero sum stochastic games,
     Operations Research Centre, University of California,
     Berkeley.

25. Schelling, T.C., (1960): The Strategy of Conflict,
     Harvard University Press, Cambridge.

26. Shapley, L.S., (1953): Stochastic games, *Proc. Nat. Ac.
     Sci.* **39**, pp 1095-1100.

27. Shubik, M., (1971): The dollar auction: A paradox in
     non-cooperative behaviour and escalation. *J. Conflict. Res.*
     15, 109-111.

28. Salter, S., (1985): Paper in this volume.

29. Takahashi, M.A., Fraser, N. and Hipel, K.W., (1984): A
     procedure for analysing hypergames. *Eur. J. Oper. Res.*
     18, pp 111-122.

30. Thomas, L.C., (1984): Games, Theory and Application,
     Ellis Horwood, Chichester, England.

31. Tucker, A.W. and Luce, R.D., (1959): Contributions to the
     Theory of Games, Vol. IV, Princeton Unviersity Press,
     Princeton.

32. von Neumann, J. and Morgenstern, O., (1944): Theory of
     games and economic behavior (2nd ed. 1947), Princeton
     University Press, Princeton.

WHY GAME THEORY "DOESN'T WORK"

K.G. Binmore
(London School of Economics)

1. INTRODUCTION

A frequently-heard complaint is that "game theory doesn't work". Such complaints reflect the gulf that exists between theoreticians with long-term scientific aims and practical men concerned with resolving immediate questions of policy. It is true that game theory as currently developed is inadequate as a predictor of human or animal behaviour except in the few special situations for which it has been tailored. Of course "meta-game analysis" is not successful in describing the Suez conflict*. Of course a representation of the Cuban missile crisis as a 2 x 2 matrix game (with Kruschev and Kennedy assumed to be rational players) is naive. But it does not follow that those who advise policy-makers can afford to neglect modern developments in game theory (notably, Harsanyi's [4] theory of games of incomplete information) or to harbour naive misconceptions about what game theory is.

Progress in game theory has been by fits and starts and numerous blind alleys have been explored and abandoned (e.g. "meta-game analysis"). Nevertheless, progress has been made. The recent successess in evolutionary biology are particularly illuminating (e.g. Maynard-Smith [7]). The difficulty in applying such new work to policy matters lies in the fact that men of affairs require answers now and in a form they find acceptable. Abstruse answers offered in the indeterminate future are therefore of little help to their advisers who must cobble together what models they can from whatever material is

---

*"Metagame analysis" (Howard [5]) seeks to explain, among other things, why some individuals choose a dominated strategy in the "Prisoner's Dilemma". Most game theorists would probably agree that the theory is no longer fashionable.

available. Game theory can only be a useful source of such
material if both its potential and its current limitations are
properly understood. This note concentrates on its limitations
under four headings: physical, behavioural, theoretical and
computational, although the third and fourth topics are treated
only cursorily. The author's prejudices will be apparent from
the fact that only peaceful applications are considered.

2. INADEQUATE PHYSICAL KNOWLEDGE

The first consideration which limits the scope for game
theory applications is the extent of our ignorance about the
physical nature of the world about us. We are ignorant across
a whole spectrum of relevant issues. We have no detailed
understanding of the capabilities and organisation of our own
bodies. Nor, at the other end of the spectrum, do we have a
detailed understanding of the institutions which form the
building blocks of our society. Certainly we have much
generalised information on these issues. We know fairly well
how a muscle works and what nerves are for. Similarly there
is no lack of learned books explaining in general terms the way
the world banking system works or the relevance of kinship
patterns among the tribesmen of New Guinea. But seldom is the
information available sufficiently sharp to admit the
possibility of a direct application of game theory.

To explain this point it is necessary first to observe that
game theoretic results usually require that we be able to
distinguish quite closely between the choices of action which
are available to a player and those which are not. It also
requires that we be able to describe, again quite closely, how
the choices made by the players interact to produce an outcome
of the game. In brief, game theory normally requires that a
precise description of the game is available.

Parlour games such as Chess, Bridge, Poker and Backgammon
present no difficulties in so far as precision of their
description is concerned. All necessary information concerning
the choices of action available to the players and the manner
in which these choices interact to produce an outcome is to be
found in the official rule-books for these games.

Games of skill like Baseball, Tennis, Soccer, Cricket or
Pool are quite different. Official rule-books exist for these
games but these do not provide all the necessary information.
In particular they say nothing about the physical limitations
of the players. But a game-theoretic analysis requires this
information and is therefore not possible unless a new rule-
book can be written which contains not only the official rules
but also all relevant information about the physical

capabilities of the players. Without such an extended rule-
book, one might well find oneself recommending or predicting
actions which the players simply cannot carry out. There would
be no point, for example, in explaining to a Junior League team
that their game would improve if they all swung their baseball
bats like Babe Ruth. Nor is the performance of a Fourth
Division soccer team likely to be improved by the observation
that they would do better if they all played like Pele. It is
necessary to know what is possible for the players and what is
not. A generalised assessment of their physical capabilities
will not be good enough unless such an assessment makes it clear
that one side has a crushing superiority over the other.
Interesting matches, that is to say matches for which a game
theoretic analysis would be of interest, will be decided by
players operating at the margin of their ability range and
presumably it will be a long time before physiology is able to
provide the data necessary to determine whether a given
standard of performance is or is not within a particular
player's ability range. This is not to say that game theory
has nothing at all to contribute to the study of games of skill.
For example, in principle game theory can be used to analyse
the home computer or video versions of such games. In these
versions, the real-life participants are simulated by simple
computer programs. For games with a considerable strategic
element, such analyses might well provide a useful source of
new ideas for a team manager. Such games, of course, are just
extensions of the table-top games currently played by football
managers or potential military commanders in planning
strategies. What we have sought to disavow is the notion that
game theory can be applied <u>directly</u> to real-life games of skill
without the availability of adequate physical knowledge of the
capabilities of the players.

So far we have considered parlour games like Chess, Bridge,
Poker or Backgammon for which the official rule-book contains a
description of all the rules that an application of game theory
requires. We have also considered games of skill like Baseball,
Tennis, Soccer, Cricket and Pool for which there seems no
possibility in the foreseeable future of producing a rule-book
adequate for game-theoretic requirements. However, the
exciting potential applications of game theory lie in the
intermediary area between these two extremes. In this area the
official rule-book is not adequate or does not exist at all.
On the other hand, there is a realistic prospect of
constructing an extended unofficial rule-book which describes
the physical and institutional constraints on the players
sufficiently precisely to make a game-theoretic analysis
feasible.

As an example, consider the game of Monopoly. An official
rule-book exists but, aside from the activities closely
specified by the official rules, there is also the important
possibility that the players may trade their properties on the
side. This possibility is admitted by the official rules but
the rules do not specify in detail the manner in which this
trading is to take place. However it is not hard to construct
some formal rules which adequately describe the manner in
which trading is carried out in practice (at least among those
who play Monopoly seriously and, as with many diversions
intended for amusement only, there are those who take Monopoly
very seriously indeed). One specification would require a
player to wait until the moment when he is about to rattle the
dice-box at the beginning of his turn, before he is allowed to
propose a trade in properties and/or money from those feasible
at that stage in the game. (To enumerate the feasible trades
in detail is a straightforward although lengthy business.) The
next step would be for the player or players whose
participation is necessary for the implementation of the
proposed trade to register their assent or dissent. If all
agree the trade takes place and otherwise not. In either case,
the proposing player then rolls the dice and the game continues.
A more elaborate version would involve counter-proposals and
counter-counter-proposals before the continuation of the game.
In this way the unofficial rules governing trade are formalised
into official rules in a larger rule-book and a formal game-
theoretic analysis becomes possible. Note that the question of
the physical capabilities of the players is not an issue in so
far as Monopoly is concerned. Nor is it an issue for the
class of economic games for which Monopoly will serve as a
paradigm. The physical requirements on a player, such as
moving counters from one square to another or passing a title
deed from hand to hand, are well within the capacity of most of
those likely to be found at the Monopoly board.

As remarked above, Monopoly is a paradigm for a large class
of real-life economic games in which the official rule-book is
inadequate but for which the players' physical capabilities are
irrelevant. These economic games do not take place in a
vacuum. Institutions have evolved which facilitate and control
different types of economic activity. By an institution we do
not only mean organisations like the Bank of England or the
Chicago Wheat Market. We also mean the process of offer and
counter-offer by means of which a carpet is sold in an oriental
bazaar or the arrangements which large companies and trade
unions employ when negotiating over wages. In order to use
game theory in this context, it is necessary in the first place
that the relevant institutions be sufficiently stable and well-
established that they can be regarded as having fixed rules of
operation which the players have no opportunity or desire to

violate. The second requirement is that the institution be
sufficiently well understood that the relevant rules of
operation can be identified and described in a precise manner.
In principle, of course, such information can be found out by
investigation. But this is by no means the same as saying
that the information is available in practice. Partly this is
because it is not necessarily obvious a priori what snippets
of information are likely to be relevant for game theory and
those who are well-informed on institutional matters are seldom
well-informed on game theory and vice versa.

In brief, there is no point in seeking to apply game theory
directly in economics unless a properly understood institutional
framework is present. As noted earlier on games of skill, this
does not mean that game theory may not be (and indeed is)
indirectly useful in the absence of hard, factual information
about institutional matters. A game-theoretic study based on
hypothetical data about the institutional framework may be
useful, for example, as a guide to an empirical worker who is
bewildered by the richness of the available data and needs
assistance is making judgements about what is likely to be
significant and hence worthy of close attention and what is
likely to be of only secondary importance.

In economics there are grounds for believing that some at
least of the important institutions are sufficiently stable and
well-established that the writing of an adequate rule-book is
not an unreasonable enterprise. An analogous statement can
also be made of certain areas in biology. In social and
political affairs however, the situation is more uncertain.
With some exceptions, notably elections and the operation of
legislative assemblies, institutions seem much more amorphous.
The fact that the negotiations over the ending of the Vietnam
war ostensibly began with prolonged discussion over the shape
of the negotiation table comes to mind. Obviously, the less
structure which can be identified in an institution, the less
likely it is that the game theory can be relevant. Possibly
however these comments say more about our ignorance of the
institutions than about the institutions themselves.

3. INADEQUATE BEHAVIOURAL KNOWLEDGE

The writing of a rule-book for a game requires a good
knowledge of the physical environment relevant to the game.
Such a rule-book will distinguish the actions the players can
choose from those which they cannot choose and explain how the
players' choices interact to produce an outcome. But the
provision of a rule-book is not sufficient in itself. To say
something about the play of a game, it is not enough to know
what actions are available. It is also necessary to know the

method the players use when choosing among the available actions.
We shall call such information "behavioural". Thus physical
information is required to determine what actions are available
while behavioural information determines how a choice is made
among these actions. Or, to put the same point slightly
differently, physical information determines what the game is
while behavioural information determines how it is played.

Ideally game theory would have access to a fully-fledged
theory of human and animal behaviour. This would tell us to
begin with what information we need in order to predict
behaviour and then provide a suitable prediction once the
information was supplied. A game theorist would then simply
have to determine the parameters of the game with which he was
involved, consult the relevant pages in his behavioural theory
textbook, and there would be a description of how the game
would be played.

Unfortunately no such theory of behaviour exists for animals
let alone for humans. However behavioural psychologists,
notably Skinner and his disciples, believe that they have the
beginnings of such a theory. They take the view that animals
and humans are best seen as stimulus-response machines. The
idea is that we cannot know what goes on in the brain of a
pigeon or a rat or another human being. We should therefore
regard the brain as a "black box" containing an inaccessible
mechanism of unknown structure which, when it receives certain
stimuli, emits certain stereotyped responses. There is much
hostility to this view. Some of this is directed with good
reason at the naivety with which the behavioural psychologists
seek to apply what is not an unreasonable basic hypothesis.
Certainly there seem no grounds whatsoever for drawing social
or political conclusions from the theory as it stands at
present. However probably more hostility is derived from the
simple fact that it is not very flattering to have to regard
oneself as a "mere" stimulus-response mechanism. Personally I
am untroubled by the thought although I have in mind a self-
image somewhat more complicated than a chocolate dispensing
machine.

Whatever a priori opinions one might hold about
behavioural psychology, it remains the case that experimental
work in laboratories does show that pigeons, rats and other
animals can usefully be described as stimulus-response machines
under certain conditions. Nor is there good reason to suppose
that humans do not behave pretty much like laboratory rats when
faced with decision problems which they have not thought about
very much. For example, if the proverbial man on the Clapham
omnibus is allowed to observe the fall of a large number of
tosses of a coin which is weighted 70% in favour of heads and

invited to predict the result of subsequent tosses, he will usually guess heads about 70% of the time and tails about 30% of the time. Precisely the same behaviour has been observed in the case of laboratory rats suitably rewarded for pressing appropriate levers. To maximise the proportion of successful guesses, one should of course guess heads every time.

As a second example, consider the following "take it or leave it" game which has been the object of some recent experimental work [1,3]. Two subjects who remain separated and totally anonymous throughout and after the game were told that a certain sum of money (sometimes quite a large sum of money) was to be divided between them provided both were willing to accept the share assigned to them. The task of assigning shares was then given to one of the subjects and this fact was known to the other subject. It was found that typical behaviour on the part of the second subject was to refuse his share (in which case both players got nothing) if the share was too small. This, of course, could well be a sound policy if the same kind of situation were liable to arise again and the actions taken by the players during the current game could be monitored by those with whom the players would be matched in the future. It would then be possible to build up a reputation for "toughness" which would be well worth having. But the experimental situation was arranged so that the second player was left with a simple choice between something or nothing, other considerations which might have been relevant under more normal circumstances having been eliminated. Nevertheless, the subjects on the whole took no note of the contrived laboratory conditions but responded machine-like to the stimuli offered them in a manner appropriate, if at all, to an entirely different environment.

To uncover behaviour like that described in the examples above requires experimental or empirical studies and, within the context of such studies, the stimulus-response machine paradigm is a natural one. But it is a mistake to be trapped into adopting a laboratory rat model of human behaviour. Such a model is only of practical use in those cases where only a small range of controllable stimuli is relevant and the process by means of which these stimuli are converted into responses is reasonably simple. But, although men (and rats too for that matter) may be "nothing more" than stimulus-response machines in some sense, it is obviously true that they cannot always be usefully modelled as simple stimulus-response machines.

Consider, for example, the behaviour of a Las Vegas casino owner in so far as the running of his Blackjack tables is concerned. Blackjack (more commonly known in Europe as Pontoon

or Vingt-et-Un) is not a complicated game. The essence of the
game is that cards are dealt to the players until they are
satisfied, the object being to obtain a hand which sums to a
total greater than that held by the dealer except that hands
which sum to more than twenty-one are "busts" - i.e. they
automatically lose. (Court cards count as ten and aces as one
or eleven.) The casino owner issues instructions to his
dealers on the strategy that the "house" is to use and the
dealers are not allowed to deviate from this strategy. The
point here is that it would be idle to suppose that empirical
studies based on the behaviour of the gambling population as a
whole when dealing Blackjack will cast much light on the nature
of the house strategy. Doubtless gamblers deal Blackjack at
home pretty much like they play it in casinos - that is to say,
with little regard to the state of their bank balances. The
same cannot be said of casino owners who would go out of
business if they made no profit. It may be that a casino
owner is a stimulus-response machine but, if so, he is one for
whom relevant stimuli include such factors as the house
strategies employed by competing casinos, the memory capacity
of inexperienced dealers and, most important, his empirical
knowledge of the way the gambling population as a whole play
Blackjack. Notice that the fact that the final piece of
information is part of the data on the basis of which the
casino owner makes his decision almost guarantees that the
manner in which the gambling population as a whole plays when
dealing Blackjack will not be a good predictor of a casino
house strategy.

   A natural comment is that the house strategies of casinos
can be discovered by an empirical study of the house strategies
of casinos. Certainly it is not possible to disagree with the
virtually tautologous statement that a sound basis for
predicting the manner in which player P plays game G is an
experimental investigation of the manner in which player P
plays game G. But there is a difference between an empirical
study and an experimental study. The use of the latter term
includes the inference that one can control the relevant
variables while no such inference is implicit in the use of the
former term. An empirical study of the house strategies of
casinos will reveal only the nature of the house strategies
under current circumstances. But for game theory purposes it
is usually just as important to know how the house strategies
will vary as the circumstances change (i.e. how the response
varies with the stimulus). Although it is sometimes possible
to devise experiments which will reveal this information, such
experiments are typically expensive and sometimes unethical.
In any case, it is seldom true that such experimental
information is available.

As it happens, few empirical studies can be so easily organised or admit such clear-cut results as a study of Las Vegas house strategies for Blackjack. The strategies are very simple with only minor differences, if at all, between casinos. However, Edward Thorpe [9], among others, has discovered the inadequacy of using such an empirical study as a model for the behaviour of the house. Using a computer, he calculated the optimal strategy for a player given the current house strategy. This is considerably more complicated than the house strategy but not so complicated that it cannot be learned by heart. The use of this optimal strategy leaves the dealer with a very slight edge over the player <u>provided</u> the cards are dealt at random. But the cards are not dealt entirely at random. In Thorpe's time, four decks were shuffled together and dealt from a "shoe". Less time was consequently wasted on shuffling and hence more time was left available for taking money from the customers. Thorpe noted, like others before him, that if a count of some sort is kept of what cards have been dealt from the shoe already, then significant statistical information about the cards remaining in the shoe will become available once a sufficient number of cards have been exposed. This statistical information can then be used to shift the bias of the game in favour of the player. However, when Thorpe sought to exploit this advantage, he found that the casinos declined to act like chocolate dispensing machines and instead reacted vigorously to the stimulus represented by his patronage of their establishments with a variety of measures (some of which make it clear that casino owners interpret the rules of Blackjack rather more widely than most of their customers). Finally, after the publication of his book "Beat the Dealer", some casinos went so far as to alter the ground rules of their Blackjack games.

Notice that a laboratory rat model will not do for a casino owner in this story. Even less will it suffice for Thorpe. Both were engaged in conscious optimising activity within a conceptual framework of some complexity. To ignore this fact and to insist on a "black box" model for their behaviour would be to abandon the structure which renders their behaviour explicable on a reasonably simple basis. The vital and fundamental point is that man is a thinking animal and not a laboratory rat or a chocolate dispensing machine. Sometimes he behaves like a laboratory rat and this behaviour needs to be studied. But to claim that the whole truth can usefully be incorporated within this viewpoint is a travesty. Not only can man reason, he can also listen or read and thereby gain access to the thought of others. In particular, he can form theories about how games are played or should be played. Alternatively he can read about such theories in books or newspapers. He may be taught such theories in school. He may

even simply note that others are more successful using
different strategies and proceed by imitation. Even where
behaviour seems at its most ratlike (e.g. in the case of
stockbrokers or merchant bankers), the players are often
acting in accordance with some outmoded theory which, in its
time, was new, revolutionary and quite inaccessible to the
mind of a laboratory rat. To repeat a much quoted remark of
John Maynard Keynes[6].

> "... the ideas of economists and political philosophers,
> both when they are right and when they are wrong, are more
> powerful than is commonly understood. Indeed the world is
> ruled by little else. Practical men, who believe
> themselves to be quite exempt from any intellectual
> influences, are usually the slaves of some defunct
> economist. Madmen in authority, who hear voices in the air,
> are distilling their wisdom from some academic scribbler of
> a few years back.... soon or late, it is ideas, not vested
> interests, which are dangerous for good or evil."

The point Keynes is making here is that human behaviour
cannot be predicted or explained independently
of the ideas which humans carry around in their heads. In
particular, a theory of human as opposed to animal behaviour
cannot hope to be widely applicable unless it takes into
account the effect which the currently accepted theories of
human behaviour have upon human behaviour. More than this, a
theory of human behaviour which does not take into account the
effect that it would have itself upon human behaviour if it
became widely accepted cannot, of necessity, become widely
accepted without contradicting its own predictions. But a
theory which could take account of the impact it would have
itself upon society would be quite remarkable. In particular
it would incorporate a dynamic model of the mechanism by means
of which humans form theories in response to external stimuli
and how they adapt or abandon these theories with the passage
of time.

The remarks above are intended to indicate the intrinsic
difficulties involved in constructing an adequate descriptive
theory of human behaviour. The stimulus-response model is
superficially attractive, like the glass slipper abandoned by
Cinderella at the ball. But the problem is an ugly sister
with a foot so immense that it is laughable to suppose that
the slipper might fit. The acceptance of this fact of life
means that experimental and empirical studies, as presently
understood, can only be expected to be useful in situations
where humans behave in a knee-jerk fashion without
introspection and only minimal calculation. Nor do there
seem grounds for anything but pessimism about the prospects

for fundamental advances in psychological theory in the
foreseeable future. How then does game theory manage given
this gloomy assessment of the state of our knowledge on the
manner in which humans make decisions?

The answer is that nearly all game theory results so far
developed postulate <u>optimising</u> behaviour on the part of the
players. It is assumed that each player is rational and well-
informed and makes an optimal choice of strategy on the
assumption that all the other players are making their choices
on the same basis. Or, to say the same thing a little
differently, a game theorist asks: What would each player do
if each player were doing as well for himself as he possibly
could? As an example, consider the familiar pass-time of
"matching pennies". By choosing his strategy in this game on
the basis of the toss of a coin, each player can guarantee that
he will break even on average. But neither player can improve
on this result given the manner in which the other player is
choosing his strategy. Thus both players are optimising given
the choice of the other.

Note that this approach short-circuits the behavioural
difficulties which we have been belabouring. We no longer ask:
How <u>do</u> people behave? We now ask: How <u>should</u> people behave?
To answer the latter question we do not need to know how clever
people are, what experience they have had or how much care and
attention they have devoted to their decisions. We only need
to be able to describe their preferences over the possible
outcomes of the game. Thus a move is good for White in Chess
independently of whether White is an international grandmaster
or a chimpanzee provided only that both the grandmaster and the
chimpanzee prefer winning to losing and regard a draw as
intermediate between these possibilities. The fact that the
chimpanzee is unlikely to think of the optimal move is
irrelevant to the question of whether the move is optimal. As
it happens, the immense complexity of Chess makes it almost
as unlikely that the grandmaster will think of the optimal move
except possibly in reasonably simple end-game positions.

Note also that a theory which postulates optimal behaviour
is one which will not invalidate itself if universally adopted.
Given that the other players behave according to the theory,
then no particular player will have any motive for deviating
from the theory because its recommendation is already optimal
for him.

The assumption of optimising behaviour makes it possible to
analyse games in a precise, no-nonsense fashion. This is very
satisfying. It is also satisfying that it leads to a theory
which is not self-invalidating. But it is of very great

importance not to lose track of the fact that this satisfaction
is bought at a price. What is lost is that the players are no
longer real people. Homo sapiens is replaced by homo
economicus. Or, if mathematicians will forgive the pun, "real
man" is replaced by "rational man". Homo economicus (or
rational man) is an ideal person invented by economists to
model economic agents. Modelling individuals as optimisers
certainly generates a theory with considerable potential

explanatory power, and there is little doubt that situations
exist for which the explanation fits the facts. In particular,
when handling substantial sums of money on a professional
basis, economic agents do seek to optimise given their
understanding of the world around them. It is therefore not
surprising that economists have found homo economicus to be
successful as a descriptive model for economic agents in
situations which are relatively easy for the agents to
understand and to assess. It is more surprising that homo
economicus is sometimes successful as a model even when the
situation is such that the agents clearly do not have any
proper understanding of what is going on. (We offer a possible
explanation of the latter phenomenon based on evolutionary
ideas later on.) However, it is a large step from these
successful applications to the assumption sometimes made by
macro-economists that it is realistic to model a large economy
"as if" each agent were homo economicus, all with perfect
foresight and all optimising on the assumption that the others
are optimising also.

Before a predictive or even an explanatory role can be
claimed for game-theoretic results, it is necessary to have
good reasons for why the players can be viewed as homo
economicus. We need an expanation of how homo sapiens became
homo economicus. Game theory short-circuits the problem of
describing human behaviour by assuming that players optimise.
But game theory offers no explanation of how they became
optimisers. The complex and difficult problem of the dynamics
of human learning behaviour is simply assumed away. But in
seeking to apply the theory, a minimum requirement is an
awareness of what has been assumed away and at least the
outline of an explanation for why the bald assumption of the
theory is reasonable in the context of the proposed application.

There are two types of influence, not necessarily
independent of each other, which could well be relevant to an
explanation of why homo economicus might be a useful
approximation to homo sapiens in a given context. The first is
the influence of education and the second is that of evolution.

Suppose, for example, that a book were generally available
which provided a precise description of optimal play in Chess.
If a sufficiently important match was to be played, then it
would be natural for both players to educate themselves by
reading the book. Both players would then play optimally and
the book would then contain an accurate prediction of their
play. Of course, under these circumstances, Chess would be a
very dull game. Indeed, at least one of the players would
probably prefer not to play since he would know from the
beginning that it was pre-ordained that he would not win.
Fortunately for Chess enthusiasts, the prospects of such a
book being written are very slim because of the computational
enormity of the problems involved. But this is not true of
games in general and business schools which have long included
rational decision theory on their syllabuses are now beginning
to include some eduation in simple game theory. However it
remains to be seen what impact this education will have
on practical game playing even in the business sphere.

Much more important than education, at least currently, is
evolution. The "survival of the fittest" is almost a cliché in
so far as the animal world is concerned. However, modern
developments with a game theory slant [7] , as popularised by
Dawkins [2] in his book "The Selfish Gene", have added fresh
insight by emphasising that it is the genes, or gene packages,
which should be seen as the players in the survival game
rather than the animals themselves. Those gene packages which
confer survival traits on the animals which carry them, given
the current environment, are those which tend to survive
because the animals which carry them are more likely to live
long enough to reproduce. Some gene packages act like
optimisers in that the traits they bestow on the animals which
carry them are optimal in so far as the survival of the gene
package is concerned. That is to say, if a conscious choice
were to be made among the behavioural traits which the gene
package could confer upon its carrier, then a choice which
optimised the prospects of survival of the gene package would
choose the gene package which we have described as an optimiser.
Optimising gene packages will tend to survive at the expense of
non-optimising gene packages competing for the same role. Of
course, what is optimal will change over time. It will be
particularly sensitive to the current mix of gene packages in
the animal population and this will not be constant. However,
in the absence of significant mutations, one might reasonably
expect convergence to an equilibrium situation in which all
gene packages act as though they were optimisers. In such a
situation game-theoretic results will be relevant.

Biological evolution is a familiar notion. Social evolution
is less familiar but operates equally remorselessly. In
biological evolution, it is important to avoid the mistake of
thinking in terms of selection operating directly on the animal
population instead of on the gene population. It is, after all,
the genes which are the basic unit of replication. Similarly,
in social evolution, it is a mistake to think in terms of
selection operating directly on the human population instead of
on the population of ideas which humans carry around in their
heads. In social evolution, it is ideas which are the basic unit of
replication but instead of being replicated from one body to another
by biological reproduction like genes, they are replicated from
one head to another by the social processes of communication
and imitation. Here an "idea" might be a complex theory or it
might be a "rule of thumb" for achieving some purpose or it
might be little more than a learned reflex action. Dawkins
uses the term <u>meme</u> to cover all these notions.

   The analogy between biological and social evolution is
useful but can be over-worked. In particular, the processses
through which selection can operate in the meme population are
far more varied than in the gene population. It remains
possible, as in biological evolution, for meme carriers (i.e.
humans) to be eliminated from society and hence for the
opportunity for them to pass on their memes to be lost. In the
commercial world, those who go out of business usually cease to
influence those who remain in business. In polite society,
those who behave unfashionably tend not to be invited any more
to smart gatherings. In social science, usually only death
eliminates the carriers of failed theories. If memes were
transmitted from head to head simply by imitation, these
considerations would be enough to fuel the evolutionary process.
But in human societies certain meme packages, in particular the
rational or scientific idea packages, owe their success to the
fact that they keep their carriers where it matters by
providing them with a more powerful method of acquiring useful
memes than simple imitation. This naturally accelerates the
evolutionary process very considerably. But it does not
necessarily mean that an equilibrium is reached sooner, if at
all. This is because the same meme packages which regulate
meme acquisition also tend to generate new memes which
constantly disrupt the evolutionary flow. The rate of
acquisition of new scientific knowledge is commonly said to be
exponential. If a criterion existed for deciding whether a
piece of information constituted new scientific knowledge or
not, perhaps this assertion could be evaluated. Without such
a criterion, all that can genuinely be said is that the
generation of new memes seems to be very fast compared with the
generation of new gene packages in biological evolution via
mutation and sexual reproduction.

All this has implications for game theory. Where meme survival is linked with the welfare of the humans carrying the meme and when the rate of production of new memes is relatively low, one might reasonably expect convergence over time to an equilibrium situation in which the surviving memes can be seen as optimisers. The humans involved in the process need not be "conscious" of what is going on in that the "explanations" offered by the memes on which they base their behaviour need not have much basis in physical reality. An individual who practises thrift and hard work may maximise his income as a result and hence become a respected pillar of the community with a consequent boost for the memes he carries. He, on the other hand, may well practise thrift and hard work on the grounds that to do otherwise would be sinful and regard the acquisition of respect and wealth as incidental.

However, one would expect convergence to be more rapid where individuals are "conscious" optimisers and so education and evolution can act hand in hand. Suppose, for example, that, in the Blackjack story told earlier, all the participants, including the gambling population as a whole as well as the casino owners, had motivations and capabilities comparable to those of Thorpe. Then the gambling population as a whole would respond optimally to a choice of dealing strategy by the casinos which would force the casinos to respond by altering their dealing strategy to cope with the new situation. This in turn would provoke a new response from the gambling public and so on. The same type of process is familiar to economists under the name of tatônnement in the study of market behaviour. Sometimes the term is used in a narrow, specialist sense, but the basic idea is that agents adjust their bids and offers in response to the bids and offers of the other agents until equilibrium is reached. It is to this mechanism which economists very plausibly attribute the tendency of markets to balance supply and demand.

The history of Chess and Bridge provides further examples Books and magazines describe current practice in the game and, in particular, they report on the play of important matches with analytical commentaries on new developments and proposals for alternatives. The Chess and Bridge literature is widely read and has a marked influence on the way the games are played. Those who take Chess and Bridge seriously therefore have little choice but to keep abreast of what is going on and to continually modify the way they play in response. Here we see in perhaps its purest form the manner in which education and evolution can act together to drive a society in the rational direction.

This last point illustrates the long term prospects for the practical application of that part of game theory (namely most of it) which is concerned with what players ought to do in their own enlightened self-interest rather than what they actually will do. If such a theory is ever to do more than grace the dusty shelves of academic libraries, it will be as a result of a combination of education and evolution driving society in the direction of the theory. In this process, the existence of the theory itself will be an important factor. This is to say that a widely applicable theory of games would, of necessity, involve a strong element of self-prophecy in the sense that the existence of the theory itself would be partly responsible for bringing about and stabilising the events which it "predicts". (Such self-fulfilling predictions are, of course, nothing remarkable on the social scene. The advertising "hype" is a particularly blatant example but similar phenomena are only too familiar in economics.) A game theorist with hopes for the widespread practical application of his subject therefore needs to be something of an optimist (as well as an optimiser) and to hold fast to a faith that the future will see a convergence between theory and practice. He also needs to be something of a propagandist as well as a theoretician since a theory of which the world is ignorant is hardly likely to influence its evolution.

A final word on this point is necessary and this will return us to our basic theme of the current inadequacy of our knowledge of human behaviour. To be a propagandist may be necessary but a line of propaganda which claims too much for a theory is likely to be counter-productive in the long run. In particular, we have outlined above a story which is capable of explaining how it can come about that homo sapiens finds himself behaving like homo economicus. But this story is only a story and its plausibility as an hypothesis depends very much on the circumstances. Certainly its relevance cannot be taken for granted even in biology. To say that evolution would lead somewhere under certain conditions is one thing. To say that evolution has already taken us there is quite another.

We illustrate this point by considering the game of Poker. This game is nowhere near as complicated as Chess and is capable of game-theoretic analysis. Have the twin influences of evolution and education brought about a convergence between theory and practice? Briefly, the answer is no. And it is not hard to see reasons why not. In the first place, anyone who has ever played Poker will be aware that, although it is necessary to play Poker for sums of money which it will hurt to lose, the winning or losing of money is not the only or even the primary motivation of many of those who play.

There is the intrinsic excitement of gambling and this is overlaid with a considerable amount of "macho" posturing. That is to say, the players seek social payoffs from the game as well as financial payoffs. In addition, there is a constant flow of inexperienced players into the game and an experienced player interested principally in money will obviously prefer to play with these inexperienced players rather than with his peers. But a good player will not play the same against poor players as he would against other good players. In particular, he will not play the game-theoretic optimal strategy which assumes optimal play by the opponents because this will not allow him to exploit the bad play of the opposition.

For these reasons, any naive attempt to use a simple game-theoretic model as a predictor in Poker is doomed to failure. Indeed optimal play in Poker strikes many experienced players as counter-intuitive in its recommendations. It is instructive however to consider such events as the "World Poker Championship" where the arguments given against a direct application of game theory are much less compelling. Here play is much closer to that recommended by game theory particularly in regard to the level and manner in which the players bluff.

4. INADEQUATE THEORY

Most of game theory is concerned with what players ought to do in their own enlightened self-interest rather than what they will do and we have discussed at length the limitations that this involves in so far as immediate application of the theory is concerned. But further limitations exist. We do not even yet have a theory of optimising behaviour which is adequate to cope with all games and large areas of game theory remain terra incognita desperately in need of exploration.

Some of the difficulties are technical but the chief problems are philosophical in that it is not so much that we cannot answer the relevant questions as that we often do not know the right question to ask. As an example, consider the glib assertion that game theorists study what happens when each player is rational and well-informed and makes an optimal choice of strategy on the assumption that the other players are choosing on the same basis. But what does "rational" mean? What constitutes being "well-informed"? Even more troublesome is the circularity built into the assertion. To decide what is optimal for a player, we need to know what is optimal for the other players. But they in turn need to know what is optimal for the first player before they can know what is optimal for themselves. This circularity means that there is no guarantee that the assertion is meaningful. And, even when a satisfactory interpretation can be assigned to the assertion, no guarantee exists that this interpretation is unique.

The difficulties raised by this and other problems are often quite vexing. But they need to be grappled with directly if progress is to be made. Game theory, like all theories, over-simplifies and sidesteps many questions but there is no room for over-simplification or side-stepping on these theoretical issues. Indeed, the reason for over-simplification and sidestepping elsewhere is to clear the decks so that a direct assault on these questions is feasible.

5. INADEQUATE COMPUTATIONAL ABILITY

It is known that Chess has a solution. That is to say, either White has a strategy which guarantees victory or Black has such a strategy or else both White and Black simultaneously have strategies which guarantee a draw. But to know that a solution exists is not at all the same thing as knowing what the solution is. One way of finding out the solution would be to make a list of all White's strategies and all Black's strategies and then systematically to try each White strategy against each Black strategy - i.e. to play all possible games of Chess. Since Chess is a finite game, there are only a finite number of possible ways it can be played. Eventually therefore this method will uncover the solution of the game. If this guarantees victory for White, the solution will consist of a strategy for White which wins against all Black strategies. Unfortunately for this methodology, even if the process were computerised on the fastest conceivable computer, the universe as we know it would be unlikely to be still in existence before the end of the calculation. Of course, there are more efficient ways of proceeding than simply by playing all possible games. But, even when one considers efficient algorithms, estimates for the time required to compute the solutions of decision problems which are quite simple to describe are capable of making the largest numbers with which cosmologists like to juggle look miniscule by comparision. It follows that there are games whose solution will never be known. Chess is perhaps one of these. This is a fact of life and with facts of life one has to learn to live.

The fact that an optimal analysis of certain games is beyond the capacity of any conceivable computer raises another issue. Should not a computer model of a player not only take account of his physical limitations but also of his mental limitations: in particular, of the data processing capacity of his brain? Simon [8] is well-known in economics for discussing such issues under the heading of "bounded rationality". However while Simon's observations are relevant they are not in a form which make them applicable in game theory. Just as an adequate physical description of a baseball player for game theory purposes would require a knowledge of the human body complete enough to make it possible to construct a robot whose physical capabilities were indistinguishable from those of the player it was built to mimic, so an adequate mental description would require a knowledge of the human mind complete enough to make it possible to construct a computer whose mental capabilities were indistinguishable from those of the player it was built to mimic. An optimising analysis would then seek optimal <u>programs</u> for the computers built to mimic the data processing capabilities of the players in question.

Such an approach would have many advantages. Computational complexities would be greatly diminished, but the chief advantage would be the potential for coping with the difficulties raised under the heading <u>inadequate behavioural knowledge</u>. In particular, an optimal analysis for two chimpanzees playing Chess would cease to be the same as that for two international grandmasters. However, while the "<u>limited rationality</u>" approach is obviously a direction in which game theory is bound to expand in the near future, only isolated results exist at present. Not only do we not have adequate models of the human or animal mind (although, perhaps paradoxically, these seem closer than adequate models of the body), we do not yet have adequately developed mathematical simplifications for computers (for example, as "finite automata" or "Turing machines").

6. CONCLUSION

In emphasising the dangers of seeking to apply over-simplified models to complex situations for which the models are ill-suited, we have inevitably provided ammunition for those who would reject game-theoretic modelling altogether. However, when contemplating our list of ignorance and inadequacy under various headings, it would be as well to bear in mind that most theories in behavioural science are so loosely formulated that it would be hard even to know where to begin in itemising their shortcomings.

REFERENCES

1. Binmore, K., Shaked, A. and Sutton, J., "Fairness or gamesmanship in bargaining?: an experimental study", to appear in *Amer. Econ. Review.*

2. Dawkins, A., (1976), *The Selfish Gene,* Oxford University Press, Oxford.

3. Guth, W., Schmittberger, R. and Schwarze, B., (1982), "An experimental analysis of ultimatum bargaining", *J. of Economic Behaviour and Organisation* 3, 367-388.

4. Harsanyi, J., "Games of incomplete information played by
    Bayesian players, Parts I, II and III", *Management Science*
    14, 159-182, 320-334.

5. Howard, N., (1971), *Paradoxes of Rationality: Theory of
    Metagames and Political Behaviour*, MIT Press, Cambridge,
    U.S.A.

6. Keynes, J.M., (1947), *General Theory of Employment,
    Interest and Money*, MacMillan, London.

7. Maynard Smith, J., (1982), *Evolution and the Theory of
    Games*, Cambridge University Press, London.

8. Simon. H., (1979), *Models of Thought*, Yale U.P., New
    Haven.

9. Thorpe. E., (1966), *Beat the Dealer*, Random House, New
    York.

BEYOND GAME THEORY -WHERE?

P.G. Bennett
(Operational Research Group, University of Sussex)

This paper traces a continuing set of developments away from
what might be called "classical" game theory. Its aim is to
explain why these changes seemed necessary, and to outline
areas in which further work still seems needed. The paper is
in four main parts. The first offers some comments on Game
Theory itself, while the second outlines the inception and
development of an approach based around the concept of a
hypergame. The third discusses some reasons why this change in
modelling format may, on its own, be an inadequate response to
certain problems, and outlines some possible ways forward.
Finally, the fourth section offers some comments on the
possible implications of these changes in terms of modelling
philosophy and usage.

1. COMMENTS ON GAME THEORY

Anyone embarking on the study of conflict is likely to
notice a number of conflicts - or at least, pronounced
dichotomies in thinking - within the field of study itself. My
own perception of two of these served to shape the ideas that
follow. The first "conflict" may well be familiar to most
people who have looked at the research literature. On one side
are those who hold that mathematical approaches - Game Theory
among others - provide a helpful aid to understanding conflicts.
On the other are those who, to state the case in its strongest
terms, regard such models as an abomination, and have very
harsh things to say about alleged dehumanisation of conflicts,
lack of realism in mathematical models, and so on.
Historically, Game Theory in particular has remained
controversial. It has been criticised for straightforward lack
of "utility" in helping people make decisions (an early critic
on these grounds was Blackett [9]), while others have seen it
as positively malign - an argument that shows little sign of

abating (see e.g. [15] , [24]).

The second, less obvious dichotomy is one that seemed to exist <u>within</u> the broad stream of work relating to game theory , and perhaps similarly within other mathematical approaches. Here, the division was between those primarily interested in the development of mathematical theory, and those concerned with applications to real-life problems, even if only in a rough-and-ready way. Amongst the first set of contributions, one typically found a great emphasis on quantification, through the use of interval utility scales and subjective probabilities, and very strong assumptions about "rationality". The aim of the exercise could be characterised as that of finding "right answers" - unique solutions, or as near as possible to unique solutions - to mathematically well-defined problems. Amongst the second set of theorists, there was much less emphasis on quantification, and fewer assumptions about rationality. For example, there was a tendency to work with ordinal preference rankings rather than utility scales. Indeed, even assumptions such as the transitivity of preferences could be broken if necessary: one was given the impression that if that was how someone actually appeared to think, it was not up to the theorist to dismiss this as intolerably "illogical". Rather than finding unique solutions, the aim of the exercise appeared to revolve around the clarification of the conflict situation. Naturally, this included guidance as to its possible resolutions, but an understanding of the paradoxes and dilemmas that might face the various participants seemed just as important. As part of this endeavour came attempts to put ideas taken from game theory into some more easily-used format, and to develop terminology more acceptable to most decision-makers[1].

It would be unjust (to both sides) to paint this dichotomy in black and white terms, and the picture presented is of course a subjective one. Nevertheless a similar rough distinction has been drawn by others - perhaps most notably by Schelling who, in the preface to the 1980 edition of his classic "The Strategy of Conflict" [30] comments:

> In putting these essays together to make the book, I had hoped to establish an interdisciplinary field that had been variously described as "theory of bargaining", "theory of conflict" or "theory of strategy". I wanted to show that some elementary theory could be useful not only to formal theorists, but also to people concerned with practical problems. I hoped too, and I think now mistakenly, that the <u>theory of games</u> might be redirected toward applications in these several fields. With notable exceptions like Howard

Raiffa, Martin Shubik and Nigel Howard, game theorists have tended to stay instead at the mathematical frontier. The field that I had hoped would become established has continued to develop, but not explosively, and without acquiring a name of its own.

Although my own interests fall very much on one side of this divide, the aim here is not to offer one-sided criticism. The assumptions made in "classical" game theory may sometimes appear breathtaking, and would indeed be so if interpreted as statements about the way the world is. However, the mathematical theorist can reasonably retort that they are not intended as such, and that it is a perfectly good research strategy to start with artificially simplified cases. Many game theorists are as well aware as anyone else of the need to keep in mind the limitations of "game" models (a thorough resume is given by Binmore in this volume [8]). On the other side of the divide, those not working "at the mathematical frontier" run the risk of not using relevant mathematical tools that are available. We may then set about laboriously reinventing the wheel - or inventing an inferior one.

For all that, Schelling's lament still strikes a chord. As the quotation above suggests, the two dichotomies are related. A concentration on idealised mathematics within game-theoretic research has its effect on how the approach is seen from outside. As a matter of personal experience, those concerned with the practical business of conflict management and resolution do not on the whole take to Game Theory: bluntly, they don't like it, and don't perceive it as relevant. Whether or not this response is justified is another matter: it certainly exists, and the perceptions underlying the response are important in themselves.

As described below, the hypergame idea originated as a response to these dichotomies. Specifically, the aim was to design a modelling approach making some use of the logical structure provided by Game Theory, while acknowledging that parties in conflict are liable to define the world in different ways, and see different issues to be important - an alleged neglect of which was central to received criticisms of the classical theory.

The hypergame concept itself has gone through several stages of development. Before describing some of these, it is important to make a preliminary point. There are two quite different ways of using theories which - like Game Theory and its derivatives - model people's decisions. The first is descriptive use, corresponding to the classical scientific

activity of explanation and prediction. The second is decision
-aiding use, in which one is using the theory to help people
make more effective decisions. (Taking the recent miners'
strike as an example, the difference is that between using
models as an outside observer to try to predict what will
happen, and using them to try to help Mssrs. Scargill or
MacGregor or - say - someone in ACAS.) Whether the recipient of
such help is one of the warring parties, a mediator, or some
other participant, the point is - to paraphrase Marx - that the
analyst tries to affect events, not merely to reflect them.
Much of our recent work has been concerned with decision-aiding
[7], in particular with the possibilities for using game-based
methods in combination with other approaches [3], [4], [14],
[19]. This paper, however, is concerned only with descriptive
modelling, leaving aside the many additional issues raised by
attempts to influence decision-makers. By the same token, I
shall leave aside the strictly normative interpretation of
game theory, in which it tells us how perfectly rational
players should behave. This remains a perfectly legitimate use
of the theory (see Binmore, op.cit.), but our main concern is
with how one might model how people do behave, and think. This
"descriptive" path is one fraught with difficulties:
nevertheless, let us see where it might lead.

2. DEVELOPMENT OF THE HYPERGAME APPROACH

The Basic Model and Some Illustrations

To return to the basic dispute between Game Theory and its
critics, it appeared that some issues dividing the two sides
might be resolved by modelling conflicts not as a single game,
but as a set of subjective games, each expressing a given
player's beliefs about the situation. This set of games is
what constituted a hypergame, in its first and simplest form
[1]. Such a system could thus be defined as:

- a set of PLAYERS, N
    (the "interested parties" in the situation, who may be
    taken to be individuals, or groups or organisations);

- for each player p, in N,
    a set of STRATEGIES for himself (S_{pp}) and for each other
    player $q(S_{qp})$. These express the possible actions or
    policies that p believes to exist for each party, and in
    turn define the outcomes perceived by p.

    Finally, we have   an ordering for each player over these
    perceived outcomes, defining the preferences p has, and
    those he attributes to the others.

Note that we ask only for perceived preferences to be
definable, rather than using subjective utility scales.
Utilities may be introduced where appropriate: if, for example,
we are prepared to take monetary gain as the sole motivation of
certain types of player. Conversely however, even the
establishing of simple preference orders may be very difficult
when dealing with less well-defined conflicts: it may be
necessary to work with partial orderings or even with
intransitive preferences. However, let us leave this aside for
present purposes.

Remaining with the basic model defined above, it is clear
that the games perceived by any two players may differ as to
the outcomes envisaged, the preferences assigned over given
outcomes, or both. At this point, two brief illustrations may
be helpful.

As an example of differing perceptions of preferences, we
may take a simple model of an arms race fuelled by mutual
distrust. Suppose our "players", p and q, are the rulers of
two nations. Each desires peace, but is suspicious of the
other. This forms the basis of a classic "cautionary tale",
originally due to Richardson [28] and found in many "standard"
texts - e.g. Rapoport [27] . A search for "security" prompts
an arms race, in which each side sees itself as making
defensive moves in response to the threat of the other's
armaments. In its extreme form, the story ends with a "mad
dash to the launching platforms" on both sides, as each tries
to pre-empt the other's expected attack. Let us now look at
this in terms of perceived games. Being at liberty to keep
things artificially simple, we shall assign a straight choice
to each side: arm or disarm. We also suppose that both p and q
perceive these same alternatives - itself a far from trivial
assumption. P places the four possible outcomes in the
following order of preference, from "best" down to "worst":

    Mutual disarmament
    Arms lead for p (p arms, q disarms)
    Arms race
    Arms lead for q (p disarms, q arms)

Unfortunately, this is not how q sees matters. He
believes that p would most prefer an arms lead for himself: in
p's preference order as seen by q, the first two outcomes above
are reversed. Finally, suppose that the same assumptions hold
with the players reversed, as regards both q's preferences, and
p's perception of them. The result is the simple hypergame
shown in Figure 1.

"MUTUAL SUSPICION" HYPERGAME

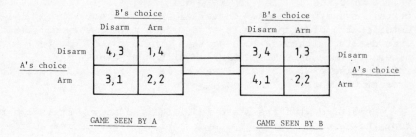

GAME SEEN BY A   GAME SEEN BY B

"PRISONERS' DILEMMA" GAME

	B's choice	
	Disarm	Arm
Disarm	3,3	1,4
Arm	4,1	2,2

A's choice

(GAME AS SEEN BY BOTH PLAYERS)

In both models, the figures in each outcome show ordinal preferences,
with more highly-preferred outcomes assigned higher numbers.
Player A's preferences are shown on the left in each case.

Fig. 1 Simple Arms Race Models

There are various ways of analysing this model. At a
strictly individual level, p perceives q to have a dominant
strategy: he sees that whatever he does, it will pay q to arm.
So p may "logically" expect q to do just that, leaving him with
a choice between an arms race and getting left behind. To
avoid his least preferred outcome, p "has to" arm. The logic
of q's game is, of course, analogous. The overall result is an
arms race nobody wanted, while each side's actions only
confirm the other's suspicions. Alternatively, suppose the two
sides engage in negotiations to stop the race: to agree a joint
move to mutual disarmament. Had there been no misperceptions,

the two sides would need only to co-ordinate their choices for
such a move to be "safe". Once agreement could be reached,
neither side would have any incentive to renege on it, and both
would realise this was so. But in the hypergame, each side
believes that the other would profit by defecting from any such
agreement, so obtaining at least a temporary advantage. So
unless there is some way of enforcing a binding agreement (and
lack of one seems undeniably "realistic"!), neither side will
"rationally" make a deal and stick to it.

It is not the intention here to argue for this model in
the context of the current East-West arms race. We have
argued elsewhere that it has some plausibility [6], but mutual
distrust is not claimed to be the only factor at work. (In his
contribution to this volume, Salter [29] suggests seven factors
inhibiting disarmament, and others will have their own lists).
Realistic or not, however, this provides an interesting
alternative to the usual game-theoretic models. The game that
would result if each side really had the preferences assumed
by the other turns out to be the famous "Prisoners' Dilemma".
Though an interesting model, this applies only to relatively
aggressive players: no element of misunderstanding is involved.
The hypergame case shows how an arms race could persist even if
each side would genuinely like nothing better than mutual
disarmament: the driving force is not any desire for
superiority as such, but rather fear of being left behind. In
this respect, our model is closer to the situation envisaged by
Richardson.

If one were looking for examples of differing beliefs
about the strategies available, an obvious set of cases are
those involving "strategic surprise" in war (or other hostile
conflicts), where one side employs a strategy which has either
been overlooked entirely by the other or dismissed as
impossible. Many such cases can be cited, and some have been
used to illustrate the hypergame idea - see e.g. [2] or [6].
However, differences in the strategies seen can also have more
subtle effects, and it may be more helpful to take a case
involving negotiations rather than just looking at the course
of hostilities. Consider, then, the prelude to the Falklands/
Malvinas War. By most accounts, this is a tale of
misperception of strategies - as well as preferences, on both
sides. Almost until the last moment, the British cabinet
apparently failed to take seriously the possibility of
Argentinian military action. Meanwhile the junta had taken
British policy to indicate relative lack of concern about the
future of the islands: hence their own actions would be hardly
likely to precipitate a war (in itself, a salutory reminder
that achieving strategic surprise is no guarantee of a happy
ending). Each side thus started by misreading the other's

concerns and - crucially - the effects its own actions would
have on the other. This much is probably non-controversial.
Here, however, we shall concentrate on the negotiations that
followed the Argentinian move onto the islands. Interpreting
these in terms of perceived games, we assume that each set of
decision-makers set out to act "rationally" given their views
of the situation.

Let us first consider the basic structure of the problem.
Bargaining and negotiation have been extensively modelled by
game theorists and others: a simple general model adapted to
the present context has the form shown at the top of Figure 2.
Each side is assigned various possible strategies, ranging
from "total climbdown" (for the Argentinians, we might define
this as an immediate withdrawal of forces and the renunciation
of any claim to the islands; for the British, an acceptance of
complete Argentinian control) to "no deal" (implying an overt
or covert refusal to negotiate on any of the issues). Between
these two extremes are various possible intermediate positions,
which can be seen in terms of willingness to accept different
proposals (for example, the British might agree to a
"leaseback" deal subject to the immediate withdrawal of
Argentinian forces). Some strategy choices on each side define
theoretical settlements, while others imply "no agreement".
For present purposes, we need not set out all possible options
and outcomes: Figure 2 simply indicates their existence
schematically. Similarly, we shall not look in detail at
possible preference structures. (Indeed, the parties may not
have clear preferences across all settlements considered,
especially where these are complex packages covering various
issues, while an important function of negotiation is also to
change preferences [20]). Without assuming that one side's
loss is necessarily the other's gain however, we shall suppose
that some outcomes would entail an almost complete climbdown
by one or other party, and that these tend to occupy the
opposing extremes of their preference rankings. This seems
plausible, especially given the internal political pressures
operating on both sides (though the need to maintain
international support provided some countervailing pressure).[2]

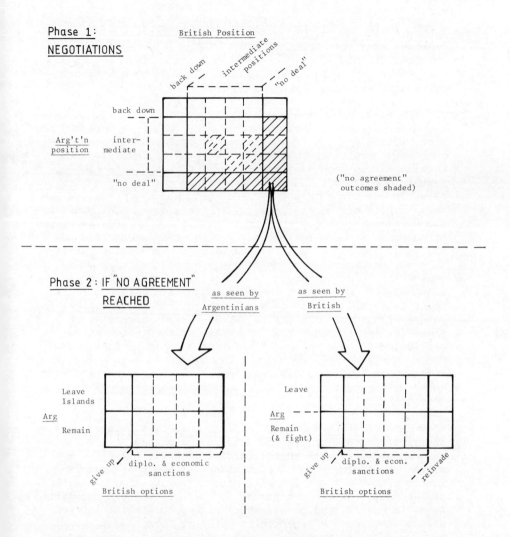

Fig. 2 Two-Phase Model of Part of Falklands/Malvinas Dispute

The basic logic of this "game" is such that any agreement can only be reached jointly by both sides. By contrast, <u>either side can unilaterally choose "no deal"</u>, leaving deadlock as the only alternative to a complete climbdown by the other. Supposing that both sides perceive a game of this basic sort, a minimum condition for the possible success of negotiations is thus that <u>there must exist some outcome perceived by both sides</u>

as feasible, and preferred by each side to "no agreement".
Such a condition may prove far from sufficient (e.g. as in the
previous arms race model). But if there is no such outcome,
agreement is "logically" unattainable since at least one side
can always guarantee itself as good a result by outright
refusal to agree to anything.

Even supposing that both sides see the situation in these
rough terms, there is still plenty of room for differences in
view. Quite apart from questions of perceived preferences,
there may be disagreement about the range of possible
agreements, as when one side is genuinely unable to meet
certain demands but cannot convince the other of this.
Arguably of most interest here however, are the perceived
consequences of "no agreement". In the simple model, this is
a convenient label. But what we have so far treated as a type
of outcome can more realistically be seen as a continuation of
the game. To model this, one can either introduce separate
subgames to describe the various stages of the problem, or
expand the original matrix (see also Thomas [32], in this
volume). We shall adopt the former approach here. Our main
point is that (at least on some interpretations of events) the
nature of the new game was perceived very differently.
Specifically, it seems that the junta started negotiations -
particularly those conducted under the auspices of Alexander
Haig - still not believing that the UK could mount a major
military operation so far from home. Observed preparations
could be discounted as a bluff and/or attempts to mollify
domestic opinion.

Extending our outline model, suppose that both sides see
the first game as before, but also perceive that if "no
agreement" is reached there, another game will result. But
what sort of game? The Argentinians believe that in this
latter phase, the British will have a variety of diplomatic
and economic options available. As before, we need not define
these in detail: the general outline appears on the left of
Figure 2. From the Argentinian point of view, "no agreement"
would result at worst in economic and diplomatic pressure.
The effectiveness of this would depend on the responses of the
many others who would become involved in that continuing game,
but would doubtless fade with time. Referring back to the
original game, we can reasonably suppose that such a result
would be preferable to anything other than a settlement very
favourable to Argentina.

The British, however, believed themselves quite capable
of going to war: for them, "reinvade" is definitely an option.
This version of the continuing game is outlined on the right
of Figure 2, giving us the skeleton of a hypergame model.

Further analysis would of course require careful consideration
of the parties' perceptions of the various possible outcomes,
and their preferences for them. While perfectly possible
however, such post hoc analysis would be of limited value now.
We have done enough to illustrate a basic point. Both sides
may have been comparing possible agreements to "no agreement".
But if the available options in the next stage of the conflict
were perceived differently, the comparisions would be against
very different bases. Nor need either side be aware of this.
Looking a little further at the British side of the hypergame,
it is clear that going to war represented an outcome preferable
to many others[3]: indeed, as time went on, such a course proved
to have considerable popular support. The cabinet could afford
to insist on an (almost) complete Argentinian climbdown as an
alternative to war. To summarise this view of events: early in
the negotiations, the Argentinian government would have seen no
reason to agree to anything other than a virtual British
climbdown. Later on at least, the British cabinet would have
seen no reason to agree to anything other than a virtual
Argentinian climbdown. It is thus plausible that at no stage
was there any potential settlement appearing in both versions
of the games, and preferred by each side to continued conflict.

The model outlined here is based on one possible
interpretation of events, which may or may not find favour (for
another, see Thomas [op cit]). Nevertheless it should, like
the previous example, serve to illustrate some of the issues
that can be considered using this fairly simple approach.
Perhaps inevitably however, others can only be addressed by
taking our ideas further. The rest of the paper examines some
of these.

Taking Subjectivity Seriously

Over the past few years, the hypergame idea has been
extended and modified in various ways [2]. One, due largely to
work by Nigel Howard, involved expanding the framework to look
more systematically at higher-order perceptions - i.e. the
players' beliefs about each other's beliefs, their beliefs
about those beliefs, and so on. It proved possible to develop
some elementary theory about the possible effects of such
beliefs, to show what properties of the system would depend on
which levels of perception.

It should be stressed that the basic idea of paying
attention to perceptions - at least, of preferences - was not
new. This had been mentioned as an area warranting future
research in Von Neumann and Morgenstern's original treatise [33],
and substantive published work dates back at least to 1956,

with a paper on misperceived preferences by Luce and Adams
[23]. Developments within the more mathematical stream of game
theory had continued, most notably with the work of Harsanyi
[16] and others on games with incomplete information. On the
whole though, these theoretical advances seem to have had
little impact on those concerned with the application of game
models (at least in areas other than in economics). The work
of Brams and his colleagues - e.g. [11] - on the effects of
deception in simple games is also of obvious relevance. Though
the relationships between these strands of research are
interesting, more far-reaching questions are raised by the need
to "take subjectivity seriously". The sort of approach
discussed so far fails to touch on some possibilities that can
arise if games are taken to be subjectively-defined in all
respects. Furthermore, these theoretical possibilities do seem
to be of importance in real life. (I shall discuss these
inadequacies in terms of different versions of the hypergame
concept, though the comments apply with at least equal force to
the modelling of conflicts as games with incomplete
information.)

Specifically then, early versions of the hypergame concept
were guilty of an accidental dilution of the principle from
which we had started, namely that different actors are likely
to construe the world in quite different ways. While it was
allowed that different games might be seen by the players, some
hidden assumptions had crept in, to the effect that all the
parties involved saw the situation in the same basic terms.
This can most easily be seen by examining how the strategies in
the various games are assumed to be related. In the simple
model above it was assumed, having defined a perceptual game
for each player, that these were linked through the existence
of at least some strategies common to each. Taking the two-
player case, it was tacitly supposed that some of the
strategies defined for p in the game he himself sees would
also appear among the strategies for him in q's game.
Similarly, some of q's strategies would also appear in p's
game. (In formal terms, this means that the perceived strategy
spaces intersect.) Were this not to be the case, then - so far
as the model was concerned - the two players would not interact
at all: nothing one could do would have any effect on the
other. Given this notion of how the games have to link up,
there are then no problems of "translation" between them. If
p implements one of his strategies, the model tells us a
priori how this will be interpreted by q. Either the identical
strategy for p appears in q's game too, or the model implies
that his actions are not "seen" by q at all unless and until
his perceptions change. ("Strategic surprise" is modelled in
this way.)

The hidden assumption in all this is that a common
language can be used to reflect the way actions are defined
and categorised by all the various players. It may be
plausible on occasion to suppose that different actors will
conceptualise the world in similar basic terms, even if they
disagree about the feasibility of certain specific actions. In
some conflicts, the parties at least have similar backgrounds
and training - thus for example, military professionals may tend
to think in a common mental frame of strategy and manoeuvre.
But this is a very large assumption to make without further
ado. Sometimes, the parties to a conflict have radically
different views of the world, and may disagree as to what the
whole conflict is actually about. (To use a well-worn but not
unreasonable example, in a conflict between master and slave
the former may see a conflict over the proper treatment of
slaves: for the latter the issue may be that of slavery.) The
original approach fails to allow for such wholesale
differences in view. In the following section, we shall
examine this difficulty further. Having re-stated the problem
for the modeller, we briefly note an existing modification of
the hypergame idea intended to overcome it. We then argue that
this is itself inadequate in some respects, and outline some
elements of a more thoroughgoing approach.

3. FURTHER ISSUES

Subjectively-defined Strategies

To restate the problem suggested above, suppose that an
analyst looking at a conflict has formulated games reflecting
each party's conceptual framework as faithfully as possible.
These games will tend to be defined in terms of the concepts
used by the parties themselves (in order to understand what
they are up to, it is surely vital to try to do this). He may
then have a set of games defined in quite different terms,
giving a very definite problem of translation between them.
There is no way of telling by definition how actions initiated
on one side will be interpreted by others, in terms of their
own perceived games. Nor may it be satisfactory to rely on the
labelling used by the parties themselves in this respect. It
is a commonplace observation that such labelling is itself
liable to be a source of disagreement.

The question of how players will interpret each other's
actions - which may well affect their own decisions - clearly
becomes a matter for further empirical hypotheses. To describe
the interaction taking place, we need statements to the effect
that if p adopts such-and-such a strategy, as defined in p's own
terms, the resulting actions will be interpreted in such-and-

such a way by q.[4] To reflect the need for such specifications
in the model, the idea of a mapping between the various
perceptual games was introduced. As first envisaged [2], these
link each player's own strategies to the set of strategies
assigned to him in each of the other players' games. In other
words, a further element was added to the definition of a
hypergame. Having defined the games themselves, we also
specify:

- For every pair of players p, q, a set of mappings from
  S_{qq} onto S_{qp} in which each strategy in the former set
  is mapped onto a subset of S_{qp}.
  These express p's interpretation of the actions entailed by
  q's strategies.

Note that these mappings need not be one-to-one. Thus q
may differentiate between actions that are all seen by q as
examples of the same thing, or vice-versa. The possibilities
of "strategic surprise" and the like are also retained (some
strategy of p's may be mapped onto the empty set, which is
formally a subset of S_{qp}).

At first sight, this revised definition may appear to meet
the problems we had in mind. However, we would now argue that
it still fails to allow for really radical differences in
perception. So far, we have taken the set of players to be
"objectively" defined: all the parties are agreed as to who the
parties are. Furthermore, we have supposed that the players do
not disagree as to the authorship of actions. So while
allowing that p's implementation of one of his strategies might
be interpreted differently by other players, these
interpretations must still have something in common: all who
see the action attribute it to p. Both assumptions seem unduly
restrictive: let us discuss the two problems in turn.

Subjectively-defined Players

On the question of the players seen, Bryant [12] [13]
points out that in real life, the "cast list" of players
perceived is indeed liable to vary. To some extent, we can
allow for differences indirectly using the previous formalism.
(For example, if p does not recognise q as a player, we can
assign q a single, "null" strategy in p's game[5].)
Nevertheless a more thoroughgoing approach is desirable,
acknowledging that the games seen by the players are
subjectively defined in all respects. In fact, we can make the
same point about perceived players as we previously made about

perceived strategies. It seems plausible that in some
instances, different parties in a conflict conceptualise the
"players" in quite different terms - rather than merely
disagreeing as to whether a specific party should be included
or not.[6] To take a rather trite example, we can imagine an
interaction between three decision-makers on the international
stage. One sees the world in terms of a set of nation-states
competing for dominance, and interprets everything that happens
within that model. Another sees the world in terms of a
struggle between certain class interests. The third
conceptualises everything that happens in terms of conflict
between certain religions. Though these are markedly
oversimplified descriptions of any real decision-makers, the
example illustrates the basic principle. It is quite possible
that the three will interact, and that the interaction will
even be reasonably predictable. The analyst does not need to
decide which, if any, of the three world-views is "correct",
nor to construct some neutral language. Given theories about
how the world is seen on each side, he may be able to make
reasonable hypotheses of the form "if p does something he
describes as s, q will see this as an example of A doing x".

Authorship of Actions

This brings us to the second issue, that of perceived
authorship of actions. There seems no good reason in general
why this should be non-contentious. In some circumstances,
there is every reason to suppose otherwise. Radical
differences in "cast list" may lead inevitably to different
beliefs as to who is responsible for which events. However,
differing attributions of actions may arise even when all
involved see the same set of players. Thus p, q and r may all
agree that they, and only they, are in the "game". But it is
still possible for some action by p to be seen by q as
emanating from r. Indeed, it may be very much in p's interests
to foster such a perception on q's part. Many classic forms of
deception seem to take this form: the work of agents
provocateurs, the "manufactured border incident", the third
party sowing discord between allies, or even the common
criminal trying to divert suspicion. Equally though,
confusions over cast lists and attributions may occur
unintentionally, perhaps without any of the parties being aware
of them. Bowen [10] provides a neat example, in the context of
some management-union negotiations. Three unions were
involved, and had formed a joint negotiating team. The
management board ultimately had certain responsibilities to the
government, but was reasonably free to negotiate on its own
behalf. Struck by "the apparent irrelevance of responses, and
the awareness of the participants that there was some hidden

barrier to understanding", Bowen hypothesised that:

> "The difficulty stemmed from what was meant and
> understood by 'we' and 'you'. If the board said 'we',
> the unions heard 'we, the government'. If a trade
> unionist said 'we', he referred to his union, but the
> board heard 'we, the joint negotiating team'. 'You' had
> reverse implications, different for each side."

A Revised Formalism

In response to the last two points, the formal model can
be generalised in two steps. Firstly, let us temporarily
suppose that the cast list is unambiguously definable, so that
one can still start by specifying a single set of players (N).
Having defined perceived strategies and preferences as before,
we can then allow for differences in perceived authorship of
actions by a straightforward modification to the previous
notion of linkage between the strategies seen. We now specify:

- For each pair of players p, q in N, a set of mappings
  from S_{qq} onto S_{Kp} where K is a subset of N, and where
  each strategy in S_{qq} is mapped onto a subset of S_{Kp}. These
  express p's interpretation of the actions entailed by q's
  strategies, in terms of the strategies he perceives for
  the set of players K.

In other words, p may interpret actions undertaken by q
as emanating from anyone in N, be it a single player (perhaps
q) or several players acting in concert.

To allow additionally for wholesale differences in how
the players are defined, a more radical revision is needed.
We are looking for a model in which the perceived games can be
entirely different, while still linked in that the
implementation of strategies in one may have some effect in
another. (Otherwise, as before, there is no interaction
between the decisions made on each side.) While a full
discussion lies beyond the scope of this paper, we can outline
the main features of such a generalised model.

Formally, we may start with a set of games as seen by
some quite arbitrary set of people, each of whom may see quite
different players. To generate these, we specify:

- A set X
denoting those whose beliefs are to be modelled:

- For every p in X, a perceptual game comprising:

a set of perceived players N_p, (the "cast list" seen
by p), and

for every q in N_p,
a set of strategies (S_{qp}) assigned by
p to q, and an ordering relationship
over the resulting outcomes expressing
the preferences p attributes to q.

So far, we have placed no restrictions on the set X. In
practice, one is likely to be interested in just those
"observers" who are also significant actors in the system. In
effect X then represents the cast list taken to be relevant by
the analyst. Even then, the set X need in principle bear no
similarity to the cast lists seen by any of its members. One
obvious restriction to consider, however, is to cases where
each "p" is at least a player in the game he himself perceives
- i.e., appears in N_p. With another simple generalisation,
mappings of the form given above can then provide a notion of
linkage between the games. It is already allowed that p may
see q's actions as emanating from some other player or
players: we now require the latter to come from a cast list
that is itself subjectively defined by p. The necessary links
can thus be specified as:

- For every pair p, q in X, a set of mappings from S_{qq} onto
$S_{K_p p}$

where K_p is a subset of N_p.
These express p's interpretation of the actions q can
take in his own game, in terms of strategies p assigns to
players within his game.

In emphasising the ways in which perceptions can differ,
our claim is not that radical differences in view will always
exist. Some conflicts - for example, certain forms of
economic competition - seem to involve no questions of
ideology or world-view: all involved may have a basically
similar view of what is going on. A game (with complete or
incomplete information) or a simple hypergame may then be a

perfectly appropriate model. The claim is rather that radical
differences in how the problem is defined may exist, and that
the models we use should allow for these possibilities rather
than defining them away. As a general principle, we would
argue for a presumption of difference between world views.
This position is based not just on observations of particular
conflicts in which various differences seem to have occurred,
but also on a general view of man as an active interpreter of
the world, constructing his own theories to guide his actions.
This ties in closely with the "personal construct psychology"
of Kelly [21]. We also wish to take on board the philosophical
point that others' constructions of the world cannot be
logically deduced from observations of their behaviour (Quine
[25]): statements about perceptual worlds thus have the status
of hypotheses, not observable facts. Our aim has been to
carry these principles as far as possible while still using
game-theoretic concepts to provide a basic modelling structure.

Dynamic Models

    Finally, our arguments so far have been stated in terms
of static models, in which all the relevant parameters are
specified once-and-for-all, and where each player's decisions
are treated as a single choice between strategies taking him
right the way through to a final outcome. Clearly however,
people's beliefs about a situation - not to mention their own
aims - are liable to change over time, so that it would be
more satisfactory to have dynamic models available. Though a
difficult area, this is one that is attracting increasing
attention. In the longer term, one might hope to combine
dynamic modelling with the thoroughgoing attention to
perceptions for which we have just argued. Meanwhile, without
claiming to offer anything like a complete specification of
how one might proceed, the following comments appear relevant.

    In terms of the framework provided by hypergames, models
can be "dynamic" in two senses.

    Firstly, one might continue to assume that the perceived
    games themselves remain constant, but analyse the
    resulting system move-by-move rather than just
    considering complete strategies.

    More radically, one might try to create fully dynamic
    models in which the perceived games may change, as
    regards available strategies, preferences and perhaps
    even players.

    On the first point, it is noteworthy that many theorists
now argue that important features of a game are lost when a

series of moves is telescoped into a single choice of strategy,
as in the original Von Neumann and Morgenstern theory. In
other words, the "normal form" (or game matrix) does not tell
us everything of interest: we may also need to look at the
"extensive form" or game tree. A useful discussion also
surveying much of the recent literature is that of Kreps and
Wilson [22] .

Combining this concern with that for differences in
perception, we may take as our basic conceptual model a set of
perceived game <u>trees.</u> As before, the games may be quite
different while still being linked: the latter notion can also
be given a more satisfactory definition in terms of moves,
rather than strategies. Figure 3 shows a simple two-player
case. We suppose that both players, p and q are agreed that
they alone comprise the relevant "cast list": further more,
each sees a simple two-move game in which p moves first. We
now link the games by a series of mappings to express how the
moves made by one player are interpreted by the other. For
example, we may suppose that both L and C map onto L'. Thus
if p makes either of these moves, q perceives that he is now
in his left-hand sub-game. Similarly, R maps onto R'. (Using
this scheme, we can also formulate a criterion of "minimal
interconnectedness": for the two players to interact, each
must have some move which, if implemented, would materially
affect the outcomes reachable in the other's game.)

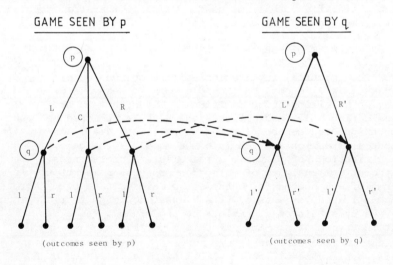

GAME SEEN BY p   GAME SEEN BY q

(outcomes seen by p)   (outcomes seen by q)

$$N_p = N_q = (p, q)$$

"Interpretation" of p's moves: L→ L' ; C→L' ; R→R'

Fig. 3 Linked Game Trees

So far of course, this still assumes that the games
themselves stay the same throughout: the players merely get to
update their positions in fixed perceptual trees as moves are
made. More generally, we may need models that are "dynamic"
in our second sense. An important set of possibilities is
opened up by allowing that the moves made at a given point may
serve to change the games subsequently seen. Thomas [op cit,
this volume] discusses the use of multi-stage games to allow
for changes in preferences: multi-stage hypergames allow an
extension of this approach. In addition however, perceptions
may change "spontaneously", or at least for reasons
unconnected with the run of play so far. We shall not,
however, attempt any discussion of such cases here. What does
seem clear is that formal theory on its own will have little
to say about such systems in general: everything will depend
on what the changes are, and whether they can be specified and
predicted.

Throughout this paper, we have been at pains to argue the
case for considering more generalised formalisms. As always,
such moves have their price. The further one goes down that
path, the less that can be proven in the way of general
mathematical theory. This brings into stark focus the
question of whether such models are then of any use at all.
In fact, however, there is a need to re-examine the whole
issue of how formal theory can be used in practice, regardless
of the specific formalisms used. We shall therefore conclude
with some suggestions on this topic.

4. ON THE ROLE OF FORMAL MODELS

The practical implications of the changes we have
suggested will depend on the role formal theory is expected to
play in modelling human decisions. Given that we are
considering only descriptive usage here, a commonsense view of
how to proceed can be pictured as in Figure 4(a). Assumptions
about a situation are fed into the model (for a game,
specifications of players, strategies and preferences).
Mathematical analysis of this then produces some output, in
terms of an outcome (or outcomes) predicted to occur. More
sophisticated variants of this view acknowledge that the input
may be uncertain or incomplete by producing a range of
alternative models. Much of this paper has concerned the
further development of formal models. However, innovations of
this sort may be of little value unless accompanied by a
re-examination of what the resulting models are actually
supposed to do. To put it bluntly, I now believe that in
practice, the "commonsense" view leads ultimately up a blind
alley, regardless of whether games, hypergames, or some yet
more sophisticated formalism is put in the central modelling "box"

Specifically, I will argue for a number of changes from such a view, grouped under two headings. In the resulting methodology, formal analysis plays a much more modest, though still useful, part.

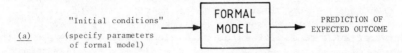

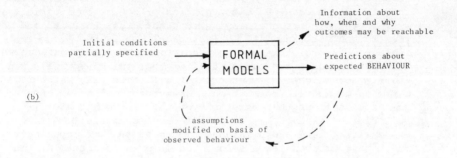

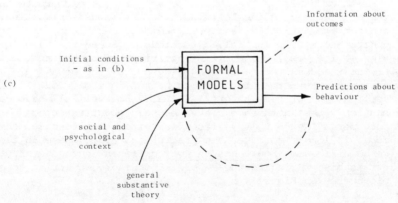

Fig. 4   Use of Models

i) Predicting behaviour

Our first suggested move is a change in emphasis from trying to predict the final <u>outcome</u> of conflicts to trying - at least as a first step - to predict parties' tactical <u>behaviour.</u> That the first aim is overambitious may be obvious

once we have differences in perception, attempts to
incorporate dynamics, and so on. But even if these extra
complications are ignored, it turns out that game models are
often only weakly predictive of outcomes. Even in simple
cases, the theory may leave many of the possible outcomes as
"solutions" - especially given the multiplicity of solution
concepts that can be used to analyse many types of game.
However such models often do have clear implications as to the
sorts of behaviour through which the players are likely to
pursue their aims. Take a very simple example, in which one
is observing a situation resembling a game of "chicken". In
this idealised case, no uncertainties or other complications
arise. The players are playing chicken: they know it, and
you (as analyst) know it. Unfortunately, this knowledge in
isolation is of little use in predicting the likely outcome.
Without more ado, one cannot tell whether one player will win
(and if so, which), or whether they will both play safe or
carry on to mutual destruction. However, examination of the
game does tell us that first player who can commit himself
publicly and irrevocably to a hard line is liable to "win".
As Schelling long ago pointed out, a good deal of apparently
odd behaviour then makes sense, and is even predictable. In
the context of a specific situation, one can thus have a good
stab at predicting the sort of tactics liable to be used,
even if one cannot yet tell who - if anyone - will come out on
top.

     An emphasis on predicting behaviour also encourages the
use of models to guide learning. As the situation unfolds,
one can monitor actual against predicted behaviour, and hence
update and modify initial assumptions. (Standing the original
view on its head, the dominant activity may become that of
trying to guess the game being played from the behaviour
observed - as certainly seems to happen in biological
applications of game theory.) Finally however, all this does
not imply that nothing can ever be said about expected
outcomes. Though this varies widely from case to case, the
model is likely to make some predictions about these in the
sense of showing how, when and why different outcomes are
attainable. Bringing these points together, we arrive at a
picture as in Figure 4(b).

ii) Use of empirical theory

     Our second set of points has to do with the sort of
information taken to be relevant when using game-based models.
If the view expressed by Figure 4(a) is accepted in its pure
form, the only empirical input regarded as relevant is that
needed to specify the formal model itself. Once we are given

the players, strategies and so on, mathematical analysis then
takes over to provide predictions - without the need to ask
any further questions. The main problem with this approach
is that one fails to seek or make use of other relevant
information and argumentation. A good case can be made for
trying to make use of at least two further sorts of input:

(a) Firstly, one should make the best possible use of what
use of whatever is known about the context under which the
game is being played. This argument again goes back to
Schelling's plea for a reorientation of game theory, in which
he argued the need to consider the "systems of communication
and enforcement" [op cit, p. 119] surrounding play, without
which the game formalism is much too abstract to tell us much.
Rather than being used in isolation, the formal model then
becomes a structure around which different strands of
reasoning can be woven.

(b) In addition to using other information about the
specific situation under study, one can go further by seeking
to develop and apply general substantive theory. While
there may be no absolute "laws of nature" available in the
field of human decision, there are at least some good rules
of thumb to describe how people tend to think and behave in
certain circumstances. To give a few examples chosen more or
less at random, one can have a good guess at how crises are
liable to affect decision-makers' patterns of thinking. We
can be reasonably sure that preferences will tend to "harden"
as the first losses in a conflict are suffered. A good deal of
work has been done to elucidate the conditions under which
strategic surprise can be accomplished, or suffered. These
fragments of substantive theory tend to be rather disjointed
and they may provide a better guide to human affairs than
relying on analysis of each situation in isolation. Far from
trying to use mathematical models to replace substantive
theory, we might thus use them as a means of expressing and
applying it.

We thus arrive at a picture as in Figure 4(c). In
general, it is not clear to what extent the so-called "common
sense view" actually reflects current modelling methods: this
probably varies a good deal. Thus some readers may find our
counter-suggestions radical, while others will regard them as
old hat. Overall, the approach to emerge is one in which
formal models provide (a) a guide to the asking of pertinent
questions about a problem in a systematic way; (b) a
convenient means of expressing the answers to such questions;
(c) a way of expressing and applying relevant pieces of
substantive theory; and occasionally (d) analysis that can

help tease out non-obvious consequences of the above. New or
not, this is a far cry from relying on mathematical
manipulation of models based only on a very abstract
specification of each class of problem.

As regards the models themselves, our earlier discussion
of formalisms should thus be read in the context of this
overall approach. One point in particular needs to be
stressed. Having argued for the formal recognition of various
complexities that can occur in real life, we might seem to be
recommending the use of very complex models - in an attempt to
capture all these features in one fell swoop. However,
nothing could be further from the truth. We have argued
elsewhere [7] that such an approach would be positively
unhelpful (in losing sight of the purpose of modelling) and is
anyway impracticable. A more modest, piecemeal approach is
advocated, in which different models are tried and compared,
and the effects of allowing for different possible
complications constantly explored. In the end, a very simple
mathematical model may turn out to be as informative as any
other ("it really does seem to boil down to a game of
chicken!). A more general framework does not necessarily lead
one to more complex models. Rather, it gives a greater range
of options as to the sorts of complexity considered at any
given point. Above all, we might hope to have some clearer
idea of what a simplified model is a simplification from.

FOOTNOTES

[1] The most widely-known of these is probably the Analysis of
Options method [18] [26], based on Howard's theory of metagames
[17]. The change in name from the original "metagame analysis",
and the dropping of other game-theoretic terminology is
significant. The method itself uses a binary format to build
up combinations of simple "yes/no" options: this allows
situations (especially those with more than two players) to be
more easily represented. The resulting system is then
analysed "in easy stages", rather than demanding a complete
specification of the game beforehand. This has considerable
practical advantages.

[2] In reality, the negotiations need to be set in the context
of a complex of related interactions affecting both preferences
and perceived options (for more on the modelling of interlinked
"games", see e.g. [7], [26]). In particular domestic pressures
seem to have constrained the extent to which either government
could "give ground" even if it wished to do so: on that basis,
there is an argument for excluding "climbdown" altogether from
the possible strategies.

3 This does not imply that the British government necessarily welcomed the conflict. Our point concerns its preferences for fighting as compared with other solutions perceived as possible, not as compared with the status quo ante.

4 It might be objected that such questions are of no consequence to the players' decisions since strategies represent single, once-and-for-all choices made at the start of the game. However, situations in which this strictly applies appear to be rather rare. We shall in any case go on to argue the case for examining the interpretation of actions on a move-by-move basis.

5 In this game, q thus appears as a "strategic dummy" (c.f. Shubik [31]), and he may also be regarded as having no stake in the situation - i.e. as being indifferent over all possible outcomes.

6 To return briefly to the Falklands example, our previous assumption that a two-player game was seen on both sides might be questioned in a way that suggests a model of the latter type. Perhaps one could interpret the Argentinian view in terms of a two-player problem, while supposing that the British government regarded the island community as another player in its own right. This would certainly fit in with public pronouncements made by both sides, which may or may not be regarded as credible.

REFERENCES

1. Bennett, P.G., (1971) Toward a Theory of Hypergames
    Omega **5**, 749-751.

2. Bennett, P.G., (1980) Hypergames: Developing a Model of
    Conflict *Futures* **12**, 589-507.

3. Bennett, P.G., (1985) On Linking Approaches to Decision-
    Aiding *J. Opl. Res. Soc.*, **36**, 659-669.

4. Bennett, P.G. and Cropper, S.A., (1985) Town Hall Maps,
    Games and things in-between, O.R. Society Annual Conference
    1984 revised version 1985, University of Sussex OR Group.

5. Bennett, P.G. and Dando, M.R., (1979) Complex Strategic
    Analysis: A Hypergame Study of the Fall of France *J. Opl.
    Res. Soc.* **30**, 23-32.

6.   Bennett, P.G. and Dando, M.R., (1982)   The Arms Race as a
     Hypergame *Futures* **14**, 293-306:   also in Dando and Newman
     (eds), *Nuclear Deterrence,* Castle House Press.

7.   Bennett, P.G., and Huxham, C.S., (1982) Hypergames and
     what they do: A 'Soft OR' Approach *J. Opl. Res. Soc.* **33**,
     41-50

8.   Binmore, K., (1985)   Why Game Theory 'Doesn't Work' this
     volume.

9.   Blackett, P.M.S., (1962)   A Critique of Some Contemporary
     Defence Thinking *Encounter,* reprinted in *Studies of War,*
     Oliver and Boyd, 1962.

10. Bowen, K.C., (1981)   A Conflict Approach to the Modelling
     of Problems of and in Organisations, in *Operational Research
     '81,* J.P. Brans (ed), North-Holland 1981.

11. Brams, S., (1977)   Deception in 2 x 2 Games *Jnl. of
     Peace Science* **2**, 171-203.

12. Bryant, J.W., (1983) Hypermaps: a Representation of
     Perceptions in Conflicts *Omega* **11**, 575-586.

13. Bryant, J.W., (1984)   Modelling Alternative Realities in
     Conflict and Negotiation *J. Opl. Res. Soc.* **35**, 985-994.

14. Cropper, S.A. and Bennett, P.G., (1985)   Testing Times:
     Dilemmas in a Action Research Project *Interfaces* **15**, 71-80.

15. Green, P., (1966)   *Deadly Logic* Ohio State University
     Press.

16. Harsanyi, J.C., (1968)   Games with Incomplete Information
     Played by 'Bayesian' Players *Management Science (A)* (i) 159;
     (ii) 320;   (iii) 486.

17. Howard, N., (1971) *Paradoxes of Rationality: Theory of
     Metagames and Political Behaviour* M.I.T. Press.

18. Howard, N., (1975) The Analysis of Options in Business
     Problems *INFOR* **13**, 48-67.

19. Huxham, C.S. and Bennett, P.G., (1985)   Floating Ideas:
     an Experiment in Enhancing Hypergames with Maps *Omega* **13**,
     331-347.

20. Ikle, F.C. and Leites, N., (1964) Negotiation: a Device
     for Modifying Utilities in: *Game Theory and Related
     Approaches to Social Behaviour* (ed) M. Shubik, Kreiger
     Publishing, Huntington, 1964.

21. Kelly, G.A., (1963) *A Theory of Personality: The
     Psychology of Personal Constructs* W.W. Norton & Co.

22. Kreps, D.M. and Wilson, R., (1982) Sequential Equilibria
     Econometrica **50**, 863-894.

23. Luce, R.D. and Adams, E.W., (1956) The Determination of
     Subjective Characteristic Functions in Games with
     Misperceived Payoff Functions *Econometrica* **24**, 158-171.

24. Martin, B., (1978) The Selective Usefulness of Game
     Theory *Soc. Studies of Science* **8**, 85.

25. Quine, W.V., (1969) *Ontological Relativity and Other
     Essays* Columbia University Press.

26. Radford, K.J., (1980) *Strategic Planning: an Analytical
     Approach* Reston Publishing Co.

27. Rapoport, A., (1960) *Fights, Games and Debates* University
     of Michigan Press

28. Richardson, L.F., (1939) Generalised Foreign Policy
     Brit. Jnl of Psychology Monographs Supplements, **23**

29. Salter, S.H., (1985) Some Ideas to Help the Arms Race
     (this volume)

30. Schelling, T.C., (1980) *The Strategy of Conflict*
     Harvard University Press, (1960, new ed'n 1980).

31. Shubik, M., (1984) *Game Theory in the Social Sciences:
     Concepts and Solutions* M.I.T. Press.

32. Thomas, L.C., (1985) Using Game Theory and its Extensions
     to Model Conflict (this volume).

33. Von Neumann, J., and Morgenstern, O., (1944) *Theory of
     Games and Economic Behaviour* Princetown University Press
     (1st Edition).

ON THE PSYCHOLOGY OF A HIJACKER

E.C. Zeeman

(Mathematics Institute, University of Warwick)

1.   INTRODUCTION

We shall use the mathematical tools of game theory [3],
decision theory [6,12], and catastrophe theory [7,10] to model
the psychology of a hijacker. A similar model can be applied
to any conflict situation involving negotiations, be it
military, political, industrial or social. We shall confine
ourselves to the case of a hijacker, however, in order to
simplify the description and highlight the essential features
of the model.

The hijack situation can be modelled initially by game
theory [3]. Although game theory explains why both sides will
want to negotiate it does not describe the subsequent evolution
of negotiations, and to understand the latter it is
necessary to analyse the conflict at a slightly deeper level.
For this we use Bayesian decision theory to model the choice
between the options open to the hijacker at any given moment.
His beliefs about the various possible outcomes of the hijack
are integrated against his preferences, giving a risk function
that can then be minimised to determine his attitude.

At a still deeper level the beliefs and preferences of the
hijacker may depend upon several parameters. For instance
examples of parameters that vary with time, and which may cause
changes in his attitude, include increasing frustration,
increasing pressure, increasing exhaustion, and increasing
rapport with the security forces. Introducing parameters into
a risk function automatically gives rise to a catastrophe model,
which shows how the attitude depends upon the parameters, and
indicates the possible slow evolutions and sudden switches of
attitude that may occur as the parameters vary.

For example, gradually increasing the pressure may cause the
hijacker to suddenly either (a) blow up the plane or
(b) surrender.   More subtly, the catastrophe model may reveal
hidden potential changes taking place behind the scenes.   For
example gradually increasing exhaustion may bias the hijacker
from (a) towards (b).   If this were the case, and if the latter
could be detected by appropriate analysis of the dialogue
between the hijacker and the security forces, then it behoves
the latter to delay increasing the pressure on him until after
the bias had taken place, in order to avoid blowing up the
hostages.

2.   GAME THEORY

We begin by recommending to the reader Michael Laver's
penetrating and entertaining book The Crime Game [3] , in which
he describes many game theoretic negotiations between criminals
and their victims.   In particular in Chapter 6 he deals with
the hijack situation, as follows.

The hijacker (or a team of hijackers) is assumed to have
hijacked a plane full of hostages, and is holding the victims
to ransom by threatening to blow up the plane with himself and
all the hostages inside.   Here the victim is the government
(or the airline or the security forces) and the ransom is
usually a complicated package involving money, escape, and
political objectives such as the release of some political
prisoners who are comrades of the hijacker.   As a first
approximation to the situation Laver proposes the following
simple 2x2 game:

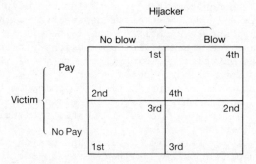

Fig. 1   Laver's hijacker game [3]

Here the hijacker has two options, either blow up the plane
or not, and the victim also has two options, either pay the
ransom or not, and so the game has four possible outcomes
represented by the four boxes shown in Figure 1.   The hijacker's
order of preference is shown at the top right corner of each
box, while the victim's order of preference is shown at the
bottom left corner of each box.   The hijacker's first choice is

pay/no blow, and his second choice is *no pay/blow,* which he
threatens to do if the victim refuses to pay. His third
choice is the total failure of the hijack, *no pay/no blow,* and
his last choice is *pay/blow* because this would involve an
unnecessary sacrifice of his own life after having achieved the
objectives of the hijack.

Meanwhile the victim's first choice is *no pay/no blow,* and
his second choice will be *pay/no blow* provided that he is
willing to pay the ransom if necessary to save the hostages.
His third choice must be *no pay/blow,* because his last choice
of *pay/blow* would involve the double disaster of losing both
ransom and hostages.

Now what does the game theory tell us? Notice first that
the hijacker has no dominant strategy: if the victim pays then
he prefers *no blow,* but if the victim refuses to pay then he
prefers *blow.* Therefore the hijacker does not want to make
the first move, but would rather wait and see what the victim
is going to do. In other words the hijacker wants to
negotiate.

By contrast the victim does have a dominant strategy, *no
pay,* because his 1st choice is preferable to his 2nd, and his
3rd choice is preferable to his 4th. However, were he to make
the first move by playing his dominant strategy and informing
the hijacker that he is definitely not going to pay, then this
would automatically lead to his 3rd choice of *no pay/blow.*
Since he prefers his 1st and 2nd choices he decides not to
play his dominant strategy after all, and tries negotiating
instead. Therefore the game theory tells us that both sides
will want to negotiate.

Laver admits that not all hijackers will necessarily have
this order of preference, and some may be so attached to
securing the ransom at all costs, that they will put *pay/blow*
ahead of *no pay/no blow,* or possibly even as high as their
2nd choice. But this does not affect the main conclusion that
they have no dominant strategy, and therefore will want to
negotiate. Similarly the victim may be sufficiently tough-
minded as to resist paying the ransom at all costs, preferring
no pay/blow to *pay/no blow* in order to support long term
deterrence against hijacking. But as before playing his
dominant strategy will immediately deprive him of his 1st
choice, and so again he tries negotiating instead.

How do we model the negotiations? One can elaborate the 2x2 game into an algorithm of metagames, by allowing each player to make a series of hypothetical moves in response to the other, but such elaboration is less convincing than the simplicity of the original game. As Thom [8] points out, any mathematical model of a piece of nature has an area of contact with reality, within which it is valid, and within which it may, subject to its own limitations, be convincing; but if it strays too far away from that area of contact then the model becomes unglued from reality, and begins to develop a life of its own, interesting possibly to the specialist, but more related to fantasy than to the application. Laver's 2x2 game above is convincing because of its simplicity, but it is too oversimplified to warrant the additional complexity of the induced metagames, without introducing further data.

Returning to our hijacker, when the negotiations begin to get serious the victim does not call in a game theorist to advise him; he calls in a psychologist. Similarly, if we wish to understand the negotiations at a deeper level we must get under the hijacker's skin and model the psychology of his attitudes and decision making processes.

3. DECISION THEORY

At any given moment the hijacker will have a great many things on his mind. His head will be buzzing with hopes and fears and aggression, and the extra adrenalin sloshing around in his bloodstream will put him on edge, and make it more difficult to think clearly about tactics and strategies. There will be a tendency to adopt instinctive attitudes, to react spontaneously, and possibly to think emotionally rather than rationally. Besides trying to achieve his main objectives and carry out preconceived plans, he will also have to continually adapt to unexpected contingencies and be constantly on the watch for countermoves by the victim. In between moments of decision and action there will be anxious and frustrating periods of waiting.

What sort of decisions does the hijacker have to make? Most of his decisions will be about what to do or say to the hostages and the victim in order to persuade them to do what he wants, such as refuelling the plane, mending a burst tyre perhaps, or getting the released prisoners to the plane by such and such time, etc. His actions and words will be coloured by the prevailing level of aggression in his attitude.

So how does one set about modelling his decision making? We suggest that it is important to focus attention upon his attitude, or more precisely on the level of aggression in his

attitude. As Allport [1] says, attitude is a mental state
that predisposes behaviour. Attitude is not only relatively
simple to describe, but also fundamental in the sense that it
is determined by the input complexity, and determines the
output complexity. Here by the input complexity we mean the
hijacker's awareness of all the possible outcomes of the hijack,
his beliefs about what is most likely to happen and his order
of preferences for what he would like to happen, all of which
must be contributory factors towards the formation of his
attitude. By the output complexity we mean the details of his
behaviour, all his actions, words, threats and ultimatums in
trying to persuade other people to do what he wants. Attitude
is the central simplicity sandwiched in between the input and
output complexities. If we explicitly model that simplicity,
while implicitly allowing for the complexities, then the model
will be psychologically realistic.

We construct the model as follows [as in 6, 11, 12, 13]. Let
x denote the hijacker's *attitude,* or more precisely the level
of aggresion in his attitude at any given moment. We assume
that x lies in a one-dimensional spectrum X, and describe the
attitude at various points of X by indicating a typical action
that might result from that attitude, as follows.

$$
\begin{array}{l}
\text{blow up the plane} \\
\text{kill a hostage} \\
\text{issue an ultimatum} \\
\text{threaten} \\
\text{negotiate} \\
\text{withdraw an ultimatum} \\
\text{release some hostages} \\
\text{surrender}
\end{array}
$$

aggression
X

We now construct a risk function R on X to represent the
input complexity. Let Y denote the *set of possible future
outcomes* of the hijack. In the game described in the last
section we assumed that Y consisted of 4 possible outcomes,
represented by the 4 boxes in Figure 1, but here Y is allowed to
be as complicated as necessary. For instance in some possible
outcomes only part of the ransom might be paid. It might become
clear to the hijacker that the victim was not prepared to budge
on the matter of political prisoners, but might be willing to
settle for money, an escape route, and an undertaking to
publicise the hijacker's cause in return for the release of
hostages. Other possible outcomes might involve the killing of
a few hostages or a shoot-out with the security forces. We shall
allow Y to be as complicated a set of possible future outcomes
as may exist in the imagination of the hijacker at the
particular moment in question.

We now introduce a probability distribution P to represent his *beliefs* about Y. Let $P(x,y)$ denote the probability of future outcome y given that he adopts attitude x now (according to the hijacker's belief). Since P is a probability distribution,

$$\sum_{y \in Y} P(x,y) = 1, \quad \text{for each } x \in X.$$

We next introduce a loss function L to represent the hijacker's *preferences* for the various outcomes. Let $L(x,y)$ denote the loss he will incur if he adopts attitude x now and y is the subsequent outcome. By "loss" we do not mean financial loss, but rather a valuation on some scale that indicates the ordering of his preferences. In the game in the last section we only used an ordering, but here we use a valuation, which is stronger than an ordering in the sense that it can indicate which choices are strongly preferred, or strongly rejected, and which are much of a muchness.

We can now calculate the *risk* $R(X)$ of adopting an attitude x. Define

$$R(x) = \sum_{y \in Y} L(x,y) P(x,y).$$

In other words the risk is the sum of possible losses, each weighted according to its probability. Define the *Bayesian decision* to be that attitude x carrying the least risk.

Having set up the model, there are now two possible ways to use it. Firstly, we could try and estimate $P(x,y)$ and $L(x,y)$ for each value of x and y, and hence compute the risk function R, and identify its minimum. This is probably what is going on subconsciously inside the hijacker's brain all the time, and may be the underlying mechanism responsible for his "instinctive" attitude at any given moment. However, it would be well nigh impossible for an outside observer to make all the estimates sufficiently accurately as to be able to compute where the minimum is with any reliability, and so it is probably better to leave most of this input complexity implicit rather than explicit.

The second approach, which is the one that we shall adopt, is to make a direct hypothesis about the shape of R that is compatible with the hijacker's known beliefs and preferences, and to predict the attitude and behaviour resulting from that shape. In other words, we make a hypothesis about the central simplicity in order to gain insight into the surrounding

complexity. Our hypotheses are given in Figure 2, and show
the changing shape of R during three successive phases of the
hijack, as we shall now explain.

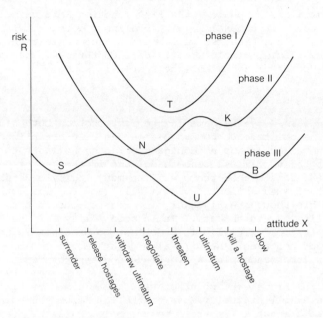

Fig. 2   Three phases of the risk function. For visual
         convenience we have drawn the three graphs below one
         another. The most important features of each graph are
         the positions and relative levels of the minima.

Phase I

During the initial phase R has a unique minimum T
representing an attitude in which the hijacker is prepared to
threaten the victim. We justify this hypothesis as follows.
At this stage the hijacker's main preference is to secure the
ransom in exchange for the hostages, and he believes there is
a good chance of success provided he can threaten convincingly.
For his threats to be effective it is important to establish
credibility; it may be risky to issue an ultimatum too soon,
lest he has to withdraw it later and thereby lose credibility.
Killing a hostage at this stage is even more risky, because
it reduces the number of hostages that can be exchanged for the
ransom, and may push the victim towards opting for a shoot-out
instead of paying the ransom. Going the other way is also
risky, because if he starts negotiating too soon this might
reduce the credibility of his threats. The more reasonable
and placatory the hijacker appears to be initially, the more

confident the victim grows about not having to pay the ransom
and the less likely will be the hijacker's preferred outcome;
consequently the more risky will be his attitude.

Summarising, in phase I the risk function R is unimodal, and
the minimum T is compatible with the input complexity.
Consequently the hijacker will begin his communication with the
victim by issuing simple clearly defined threats, such as
blowing up the plane unless the ransom is paid.

Phase II

As explained by the game in the last section, both sides will
want to negotiate once communication has been established.
Since the hijacker has no dominant strategy he wants to find
out what the victim is going to do, and if the latter agrees
to pay the ransom then there will be many logistic details to
be settled. Meanwhile the victim will want to find out as much
as possible about the hijacker, in order to assess the
credibility of his threats. It is good policy for the victim
to play it cool initially in order not to precipitate any
killing of hostages; he will also gain time to assemble his
security forces and plan appropriate countermeasures.

From the hijacker's point of view the minimum of his risk
function R will gradually move to the left from the point T in
the phase I graph in Figure 2 to the point N in the phase II
graph, representing an attitude in which he is prepared to
negotiate. We suggest that a consequence of this move to the
left will be the emergence of another higher minimum K on the
right, counterbalancing the move to the left. It is not as if
there will be any reduction in the total amount of aggression
felt by the hijacker, but rather a splitting of it into the
bimodality of N plus K. It is diplomatic for him to move from
threatening to negotiating in phase II, and less risky in the
precise sense of the model, in that the negotiating attitude is
more likely to ultimately achieve his preferred outcome. At
the same time, however, there will be a hardening at the back
of his mind in case the negotiations are unsuccessful, and the
new minimum K represents an attitude in which he is quite
prepared to kill a hostage if necessary. If the victim
exploits the hijacker's apparent softening by prevaricating,
then the hijacker may well punish him by killing a hostage in
order to concentrate the victim's mind. For instance in the
1977 Mogadishu hijacking [3], the hijacker shot the pilot in
cold blood as a punishment and warning to the security forces
against further prevarication.

We explain the nature of the change of attitude as follows. As the victim's intransigence gradually dawns upon the hijacker, the latter's attitude towards negotiating becomes more risky in the strict sense of the model, because his belief that this will eventually lead to the paying of the ransom begins to dwindle. Meanwhile his attitude towards killing a hostage becomes relatively less risky, because he begins to believe that it might in fact increase the probability of the preferred outcome. This is illustrated by the sequence of graphs in Figure 3.

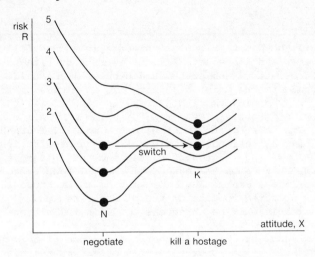

Fig. 3 Switch of attitude. The victim's intransigence causes a gradual change 1,2,3,4,5 in the risk function, and hence a sudden switch of attitude from N to K by the hijacker. For visual convenience we have drawn the graphs above one another. The important feature is the relative levels of N and K.

The minimum at N is rising, while that at K is falling relative to N. N is below K in graphs 1 and 2, level with K in graph 3, and above K in graphs 4 and 5. In fact N has coalesced with the maximum in graph 5 and is about to disappear. The global minimum of each graph representing the Bayesian decision is marked with a dot, because this is the attitude that will be adopted. At graph 3 the global minimum switches from N to K, and therefore the hijacker will suddenly switch his attitude from negotiating to killing a hostage. His calculations of the risk functions may have been either conscious or subconscious, and hence his switch of attitude may appear either as a rational choice or as an instinctive reaction, similar to a sudden switch of perception [13].

This is the essence of catastrophe theory, a continuous
change surprisingly causing a discontinuous effect. The
gradually changing risk function causes the sudden catastrophic
switch of attitude. The victim may well be caught off guard
because the negotiations seem to be proceeding smoothly and
apparently going his way when all of a sudden the hijacker
turns nasty and kills a hostage. On the other hand, if the
hijacker believes that the victim is negotiating seriously
then N will remain below K, and there will be no need to kill
a hostage.

Phase III

Even if the negotiations are proceeding successfully, both
the hijacker and the victim will be aware that time is on the
side of the latter, and so the hijacker will gradually harden
his attitude to the point of delivering an ultimatum if
necessary. The minimum of his risk function R will therefore
gradually move to the right from the point N in the phase II
graph in Figure 2 to the point U in the phase III graph,
representing an attitude in which he is prepared to issue an
ultimatum. We suggest that a consequence of this move to the
right will be to push the second minimum K even further to
the right into a minimum B, representing an attitude in which
he is now prepared to blow up the plane if necessary. For
that is exactly what he threatens to do in the ultimatum.

We also suggest that a further consequence of this move to
the right will be the emergence of another counterbalancing
minimum S on the left, representing an attitude in which he is
prepared to release hostages or even surrender. The more
closely he has to face the possibility of failure of the
ultimatum, leading to his own imminent death when he has blown
up the plane, the more likely he is to explore the risk of
alternative options that might save his life. In effect the
amount of aggression behind N will be split bimodally between
S and U, and so the total aggression will be split trimodally
between S, U and B.

If the victim perceives the emergence of S he may seize the
opportunity to put pressure on the hijacker, for instance by
renegotiating the deadline, calling his bluff, or threatening
a shoot-out. The purpose of putting on the pressure is to
raise the level of U until it is higher than S, in the hope
that this will cause a sudden switch in the attitude of the
hijacker from U to S, resulting in his surrender. The danger
of this course of action, however, is that it might precipitate
the opposite switch from U to B, resulting in the hijacker
blowing up the plane, with himself and all the hostages in it.
Therefore it is crucial to know whether S is higher or lower

than B before putting on the pressure to raise U. The victim's best strategy may be to appear to concede to the ultimatum so as to keep the hijacker's attitude at U, while at the time entering a dialogue aimed at reducing the risk of S. For example the victim might offer to pay a modified ransom in exchange for the hostages, guaranteeing at least some benefit to the hijacker's cause. He could stress how much the hijacker has already benefited from the publicity so far, and could offer generous surrender terms emphasising the advantages of life over death. The increasing exhaustion of the hijacker over a number of of days may also contribute to the effect of lowering S relative to B, by sapping the hijacker's willpower to blow up the plane, and reducing the risk of S by modifying his beliefs and preferences.

If the victim can estimate the relative levels of S and B by carefully monitoring his dialogue with the hijacker, he may be able to detect when S drops below B, and choose that moment to put the pressure on in order to trigger the switch from U to S.

Notice that in the above discussion the catastrophes, in other words the sudden switches of attitude, have played a much more significant role than the gradual changes of attitude. This is surprising, because if you want to change someone's mind it might appear at first sight to be more natural to try and argue them gradually out of their present position. Such argument, however, can be uphill work, and is indeed literally uphill if you are working against a subconscious risk function. It may be a more effective strategy to persuade them first to lower a distant minimum and second to raise the existing minimum, for then the subconscious will do the work for you by switching the attitude automatically. If the situation is trimodal as in phase III above, the strategy may involve several steps, and to decide the best tactical order in which to make those steps it may be necessary to appeal to the higher dimensional geometry of catastrophe theory, as follows.

4. CATASTROPHE THEORY

The sophisticated theorems of catastrophe theory classify universal families of functions [5,7,10]. Given an evolution from a function R_1 to a function R_2 we can ask the question: what is the smallest universal family containing that evolution? The answer is called the unfolding of the evolution.

For example consider the evolution of the risk function shown in Figure 2 from phase I to phase II, going from unimodal to bimodal, in which T bifurcates into N plus K. The

unfolding of this evolution is the cusp catastrophe illustrated
in Figure 4.

The cusp catastrophe is a 2-dimensional family of risk
functions parametrised by the 2-dimensional parameter space C
represented by the horizontal plane in Figure 4. The attitude
spectrum X is represented by the vertical line. For each
point $\check{c} \in C$ we have a risk function R_c on X. To illustrate all
the R_c's would require a 4-dimensional picture, which is
difficult to visualise, and so we confine ourselves to the
qualitatively most important feature, namely the maxima and
minima of the R_c, as follows. Define the attitude surface A
to be the surface in the 3-dimensional space C x X given by

$$A = \{(c,x) ;\quad R_c \text{ has a maximum or minimum at } x\}.$$

Then A is the folded surface shown in Figure 4, of which the
shaded area represents maxima, and the rest minima. The sheet
at the back of A represents the attitude T (threaten), the
bottom left front sheet represents N (negotiate), and the top
right front sheet represents K (kill a hostage). The two
front sheets merge together at the back, but form separate
layers at the front. The fold curve $\bar{Q}$ of A separates the
maxima and minima and projects down onto the cusp Q in C.

Q separates C into unimodal and bimodal risk functions as
follows. If the parameter point c lies outside Q then the
corresponding risk function R_c is unimodal, and so there is
a unique point of A above c corresponding to the unique
minimum of R_c. If c lies inside Q then R_c is bimodal, and so
there are three points of A above c, corresponding to the two
minima N and K separated by a maximum. The dashed line L
bisecting Q is called the Maxwell line [10], and is the set of
points c for which the minima N and K are at the same level, as
in graph 3 of Figure 3. If c lies to the left of L inside Q
then N is lower than K as in graphs 1 and 2 of Figure 3, so N
is the attitude that will be adopted; and if c lies to the
right of L then vice versa. Therefore L is the frontier across
which sudden switches of attitude will take place, and the
dashed parabola $\bar{L}$ is the induced frontier on the attitude
surface A above. The gradual change of risk function
illustrated in Figure 3 is represented in Figure 4 by the path
q in the parameter space C, inducing the path $\bar{q}$ in the
attitude surface A above, containing the sudden catastrophic
switch from N to K as q across L.

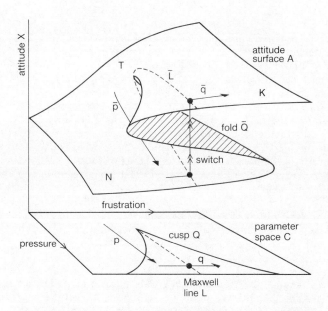

Fig. 4 Cusp catastrophe unfolding the evolution from phase I
        to phase II. Paths p,p̄ represent the gradual change
        in the hijacker's attitude from threatening to
        negotiating. Paths q,q̄ represent the catastrophic
        switch of attitude from negotiating to killing a
        hostage as q crosses the Maxwell line L.

The evolution from phase I to phase II illustrated in
Figure 2 is represented in Figure 4 by the path p in the
parameter space C, inducing the smooth path p̄ in the attitude
surface A above, representing the smooth evolution from T to
N without any sudden switches of attitude.

The two main advantages of considering the unfolding of the
evolution in Figure 4 are as follows. Firstly it provides a
comprehensive picture of the variety of smooth changes and
sudden switches of attitude that are possible under given
changes of parameter, and explains how they are all
interrelated. Understanding what is possible improves the
chances of prediction.

Secondly the knowledge that the unfolding is 2-dimensional
tells us to look for two parameters governing the risk
function, and hence governing the changes of attitude. The
clue to identifying these two parameters is given by the paths
p and q. The parameter parallel to p is called the splitting
factor (because it splits A) and that parallel to q is called

the normal factor (because it correlates with x).

The path p representing the evolution from phase I to phase II is essentially caused by the pressure of time on the hijacker. He knows that time is on the side of the victim, and that unless he swallows some of his aggression and gets down to the negotiations the situation may well drift out of control. Compare the hijacker with a kidnapper, for instance, who has his hostage stashed away in some secret hideout, and hence is under no such pressure to negotiate. By contrast the hijacker is under increasing pressure, and so we can identify pressure with the splitting factor, whether it be pressure of time or any other kind of pressure exerted by the victim.

The path q, meanwhile, is caused by increasing frustration due to the intransigence of the victim, and so we identify the normal factor with increasing frustration, as shown in Figure 4.

At this stage the reader will see that the model is closely related to the classical cusp catastrophe model of aggression [9,10,11] shown in Figure 5. The fight/flight mechanism is phylogenetically ancient and permanently wired into the brains of most animals, and is therefore a basic template for many types of animal and human behaviour. The similarity of Figures 4 and 5 shows that our hijack model is compatible with basic instinctive human behaviour.

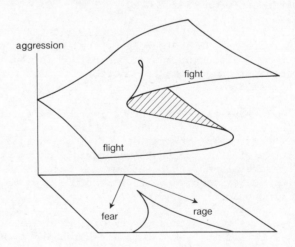

Fig. 5 Cusp catastrophe model of aggression.

Phase III:

The evolution from phase II to phase III in Figure 2, in which N bifurcates into S plus U, unfolds into another cusp. The question arises: how do these two cusps fit together? The answer is revealed by unfolding the total evolution from phase I to phase III. This gives a butterfly catastrophe [7,10], which has four parameters, namely butterfly and bias factors in addition to the splitting and normal factors that we have already identified. We suggest identifying all four factors as follows:

normal factor   : increasing frustration of the hijacker

splitting factor : increasing pressure on the hijacker

bias factor   : increasing exhaustion of the hijacker

butterfly factor : increasing rapport with the victim.

We must now justify these suggestions in terms of the geometry of the butterfly catastrophe. The butterfly is a generalisation of the cusp, and the easiest way to understand it is to see the effect of the two new parameters on the cusp. Consider first the butterfly factor.

One of the consequences of the dialogue between the hijacker and the victim over a period of possibly a few days will be an increasing rapport between them. Each learns about the other's beliefs and preferences, strengths and weaknesses. The victim's security forces take care to funnel the dialogue through a single experienced negotiator, whom the hijacker can then get to know personally, and on whose word he can begin to rely. It is this increasing rapport that essentially facilitates the bifurcation of N into S plus U, because it makes it easier for the hijacker to deliver his ultimatum to the victim through the negotiator, together with all the appropriate logistical details. The negotiator has to accept the ultimatum, or at any rate has to promise to convey it faithfully to the victim, otherwise he will lose the hijacker's confidence. At the same time the rapport helps to establish the new minimum S, because the negotiator will work at creating a perception of the surrender option in the hijacker's mind [13]. Thus the rapport is responsible for causing the trimodality of phase III. Therefore we can identify increasing rapport as the butterfly factor, which geometrically causes the evolution from Figure 4 to Figure 6, as we now explain.

Corresponding to the trimodality of phase III the attitude surface A in Figure 6 has evolved into three sheets of minima which merge together at the back but form separate layers at

the front: the top right front sheet represents B (blow), the
bottom left front sheet represents S (surrender), and the
middle triangular sheet in between represents U (ultimatum).
The cusp Q has evolved into a figure containing three cusps,
which form the triangle underneath the triangular sheet U
above. The Maxwell line L has evolved into a Y-shape,
separating the regions of C dominated by B,S and U respectively.
The path s in the parameter space C represents increasing
pressure on the hijacker at a high level of frustration, and
induces the path $\bar{s}$ on the attitude surface A above, containing
the disastrous sudden switch of attitude from U to B (blowing
up the plane) as s crosses the right branch of L. The
increasing pressure is due partly to the pressure of time on
the hijacker, but may also be due to a deliberate policy by
the victim. The parallel path t represents a similar increase
of pressure, but this time at a lower level of frustration,
and consequently induces the opposite, and more desirable,
switch from U to S (surrender) as t crosses the left branch of
L.

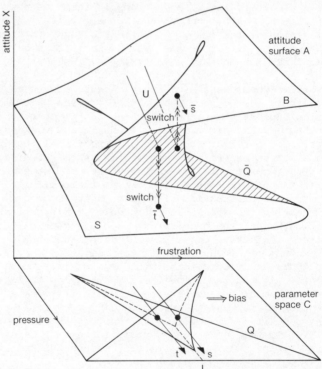

Fig. 6   Section of the butterfly catastrophe modelling phase
         III. Paths s, $\bar{s}$ represent increasing pressure
         inducing the disastrous switch from ultimatum to
         blowing up the plane. Paths t, $\bar{t}$ represent the
         opposite switch from ultimatum to surrender.

The close similarity between these two paths in C reveals
the delicacy of the situation, and emphasises the victim's
need to know the precise position of the hijacker relative
to L before he dares to increase the pressure. This is where
the bias factor comes in.

Geometrically the effect of small positive bias is to move
Q and L to the right. Therefore, relative to Q, both the
hijacker and the frustration axis move to the left, transforming
the disastrous path s into the desirable path t, while
maintaining the same level of frustration. Thus a prior
application of sufficient positive bias makes it safe for the
victim to subsequently step up the pressure in order to trigger
the hijacker's surrender without fear of blowing up the plane.
We suggest that the increasing exhaustion of the hijacker is a
positive bias factor, as explained in the last section. This
may not be the only bias factor, because for instance there may
be an overriding negative bias in the hijacker's personality;
a fanatic may prefer martyrdom to surrender however exhausted he
may be. Estimation of the bias must be one of the main tasks
of the consultant psychologist who is monitoring the dialogue
for the victim.

As we have said, the effect of small bias is to move Q and L
sideways. Geometrically the effect of large bias is to
withdraw one of the horns of Q and abolish the corresponding
branch of L, as shown in Figure 7 (for proof see [10]).

Negative bias leaves intact the right horn of Q, which is
the old cusp that originally evolved in phase II. Positive
bias leaves intact the left horn, which is the new cusp that
evolved in phase III. Figure 7 explains how these two cusps
fit together, and how they are both related to Figure 4 by
suitably biasing and applying the butterfly factor. Figure 7
also shows how the disastrous path s of suddenly blowing up the
plane is converted by positive bias into surrendering. Notice
that positive bias moves the dot on s lower: this means that
it will take more pressure to make the hijacker surrender than
it would have done to have made him blow up the plane.

For further discussion on the butterfly catastrophe, and
the effects of other possible paths in parameter space see
[2,5,10].

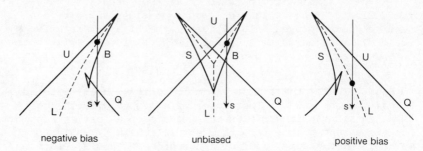

negative bias   unbiased   positive bias

Fig. 7   The effect of bias. In the unbiased situation the
        path s of increasing pressure induces the disastrous
        switch from ultimatum to blowing up the plane.
        Negative bias reinforces this switch, but positive bias
        converts it to the opposite switch of surrendering.

5.   EMOTIONAL DELAYS

Before concluding we return to the theme of emotional versus
rational decision making. So far in this paper we have
assumed that the switch of attitude takes place at the Maxwell
point, where one minimum falls below the level of another,
as in graph 3 of Figure 3. This is essentially a property of
intelligent decision making, and a mathematical consequence
of minimising functions. By contrast in non-intelligent
decision making the switches are delayed until the bifurcation
point [12]. Here a bifurcation point means a point where a
minimum coalesces with a maximum, as in graph 5 of Figure 3.
In Figures 4 and 6 the set of bifurcation points is Q. A
typical example of non-intelligent decision making is
Darwinian evolution, since it is governed locally by natural
selection [14].

We suggest that a similar delay may occur in switches of
attitude, if the latter happen to be determined by the
emotions rather than by rational thinking, as illustrated
in Figure 8 and as we explain below.

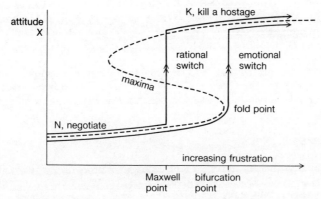

Fig. 8   Rational and emotional switches.

Emotions and moods are governed by the limbic brain [4] rather than the cortex, and changes of mood characteristically exhibit delays and hysteresis. Therefore to model the limbic brain we choose a mathematical tool with similar characteristics, such as a dynamical system, whose attractors switch at the bifurcation point. By contrast, the cortex is responsible for the visual and auditory systems, focus of attention, rational thought and intelligent response, none of which exhibits such delays as the limbic brain. Therefore to model the cortex we choose a mathematical tool with the opposite characteristics, such as a risk function, whose minima switch at the Maxwell point.

If the hijacker's attitude is primarily determined by rational thought, be it conscious or subconscious, calculating or instinctive, then the switches of attitude will take place when crossing the Maxwell line as we have assumed throughout the paper. If, however, his attitude is primarily determined by his emotions, then the switches will be qualitatively the same, but temporarily delayed until crossing the cusp line beyond the Maxwell line. The difference might be observable by monitoring the dialogue with the hijacker.

6.   FURTHER DEVELOPMENTS

What needs to be done is to analyse the recorded dialogues of various hostage situations, to see if they can be interpreted in the light of the model and classified accordingly. Could the model have given a better insight in certain cases, or made better predictions of the hijacker's change of attitude? Could it be used predictively in future hijacks? A further complexity that needs analysing is to combine the model of the hijacker with a model of the victim

into a single interacting model in higher dimensions. More
generally it would be interesting to apply a similar model
with appropriate parameters to other types of conflict
situations.

I am indebted to Peter Shapland who first suggested the
problem to me, particularly for the case of hostage-takers
in prisons, where the inmates sometimes take the warders
hostage.

REFERENCES

1. Allport, G.W., (1935) Attitudes, Handbook of social
    psychology (ed: C. Murchison), Clark University Press,
    Worcester, 789-844.

2. Callahan, J.J., (1980) Bifurcation geometry of E_6,
    Mathematical Modelling **1**, 283-309.

3. Laver, M., (1982) The crime game, Martin Robertson,
    Oxford.

4. MacLean, P.D., (1973) The triune concept of brain and
    behaviour, Hinks Memorial Lectures, Toronto University
    Press.

5. Poston, T. and Stewart, I.N., (1978) Catastrophe theory
    and its applications, Pitman, London.

6. Smith, J.Q., Harrison, P.J. and Zeeman, E.C., (1981) The
    analysis of some discontinuous decision processes, *Euro.
    J. Op. Res.* **7**, 30-43.

7. Thom, R., (1972) Structural stability and morphogenesis,
    Benjamin, New York.

8. Thom, R., (1980) The role of mathematics in present-day
    science, *Proc. Congress Philosophy of Science*, Hanover.

9. Zeeman, E.C., (1971) Geometry of catastrophe, *Times
    Literary Supplement*, 1556-7.

10. Zeeman, E.C., (1977) Catastrophe theory: selected papers
    1972-1977, Addison-Wesley, Reading, Mass.

11. Zeeman, E.C., (1980) Catastrophe models in administration,
    Proc. Assoc. Inst. Res. **3**, 9-24.

12. Zeeman, E.C., (1981) Decision Making and Evolution,
     Theory and explanation in archaeology, (eds: C. Renfrew,
     M.J. Rowlands, and B.A. Segraves), Academic Press, New York,
     315-346.

13. Zeeman, E.C., (1983) Sudden changes of perception,
     Logos et Theorie de Catastrophes, Cerisy, France.

14. Zeeman, E.C., (1984) Dynamics of Darwinian evolution,
     Colloque des Systemes Dynamiques, Foundation Louis de
     Broglie, Peyresq.

CONFLICT OF BELIEF: WHEN ADVISERS DISAGREE

S. French

(Department of Mathematics, University of Manchester)

1.   INTRODUCTION

The majority of papers at this conference deal with conflict
of preference between different actors within a system.   But
other types of conflict are possible.   Here we consider the
case where different experts give conflicting advice to a
single decision maker.   Analysis of such conflict, I will
argue, should take a very different form to the analysis of
conflict of preference.   Factors such as bargaining strength or
principles of democracy are no longer relevant.   Instead it is
necessary to consider the ideas of relative expertise, common
knowledge, honesty and so forth.

After a brief introduction to the problem, a particular
example due to Howard Raiffa is given.   The following sections
indicate how a three different approaches would 'resolve' the
conflict in the example.   There is also an appendix giving a
brief guide to the literature.   We shall suppose that a single
decision maker (DM) faces a real, well-defined decision problem,
in which the consequences of his choice depend upon the
occurrence of events in a particular (σ-)field, a.   We shall
make the further assumption that the DM is Bayesian.   Thus he
will represent his uncertainty about the events in a by a
subjective probability distribution, π, and his preferences
between consequences by a utility function, u.   We shall not
be concerned with the assessment of u ;   we shall simply take
it as given.   Our concern will be with the representation of
his uncertainty and the manner in which that representation
should change as he learns the opinions of others about the
likelihood of the various events in a.   To be precise, we shall
suppose that the DM asks n experts for their advice and that
each states his beliefs as subjective probabilities,
p_k (k = 1,2,...,n).   Here p_k is to be interpreted as a

probability mass or density function according to context.
The DM wishes to form his own probability distribution, π, in
the light of their advice. π should clearly be a function of
the p_k and arguably several other factors; but what function
and what other factors?

There are a number of points which should be clarified at
the outset. First, the 'experts' need not be expert. They are
simply people that the DM asks for advice. All that matters
about the experts is that they understand what the events in
$\mathcal{A}$ are: a necessary condition for them to be able to give
meaningful probabilities. Second, we shall allow the experts
to give advice only in terms of stating their probabilities.
Admittedly, this denies the model much realism, but it does
enable us to focus on certain issues which might be obscured in
a more general setting. Third, the experts have no
responsibility for the decision. Their role is to provide
advice. The DM has sole responsibility for the decision. Thus
we are concerned with the problem known as the expert problem.
(Other problems related to group consensus beliefs are
indicated in the bibliography).

Our approach in this paper is not to discuss the problem in
any great generality. Rather we focus attention on one
particular example, published in Raiffa [19, pp. 228-233], but
suggested originally by John Pratt. Moreover, in our
discussion of this example we concentrate on one issue:
independence preservation, viz. if the DM and the experts agree
that two events are independent, should the DM assign
independent probabilities to the events having heard the
experts opinions? However, tightly focussed though our
discussion will be, it will, I think, provide an introduction
to the complexity of the expert problem.

In the next section Raiffa/Pratt's example is stated.
Sections 3 and 4 consider the "solution" of the example by,
respectively, linear and logarithmic opinion pools. Section 5
suggests that any opinion pool, linear, logarithmic or
otherwise, is susceptible to many faults and biases and,
indeed, provides a very poor solution to the expert problem.
Section 6 introduces a solution to the problem based upon
Bayesian updating. The presentation is non-technical and, I
hope, intuitive. Mathematical details of similar models are
available elsewhere. Finally, the appendix comprises a
glossary of terms and a guide to the literature on conflict of
belief.

2. RAIFFA'S EXAMPLE

The DM faces a problem which turns on two sources of
uncertainty: the future value of the British pound and
whether or not a piece of equipment will fail. The DM
consults two experts. Both experts are adamant in believing
that any change in the value of the pound is (probabilistically)
independent of the possible failure of the equipment.
However, one expert believes that the pound is more likely to
rise than fall and that the equipment is more likely to fail
than succeed; while the other expert believes that the pound
is likely to fall and that the equipment is likely to succeed.
Their assessments are given in Table 1. We shall use the
following notation for events.

       A - the event that the British pound rises

       B - the event that the equipment succeeds.

a is the field generated by $\{A,B\}$. An over-bar indicates
complementation, e.g. $\overline{A}$ is the event that the £ falls.

Clearly the experts' beliefs are in considerable conflict.
Thus the DM, in forming his belief, has to resolve this
conflict. How should the DM form his subjective probability
distribution $\pi(.)$ given the experts probabilities $p_k(.)$,
k = 1,2,? Since the experts both believe A and B to be
independent, we might presume that $\pi(.)$ should assign
independent probabilities to A and B, particularly if we
assume, as we shall, that the DM himself believed the events
to be independent before he consulted the experts. Surely in
resolving conflict $\pi(.)$ should preserve what agreement there
is. Well, that we shall discover is both somewhat of a moot
point and a forlorn hope.

Expert 1

Equipment:	E rises A	E falls $\bar{A}$	
succeeds, B,	0.14	0.06	0.20
fails, $\bar{B}$.	0.56	0.24	0.80
	0.70	0.30	

Expert 2

Equipment:	E rises A	E falls $\bar{A}$	
succeeds, B,	0.24	0.56	0.80
fails, $\bar{B}$.	0.06	0.14	0.20
	0.30	0.70	

Table 1 The assessment of the two experts. Note that both experts assess the probabilities of A and B to be independent: viz.

$$p_k(AB) = p_k(A)p_k(B) \quad \text{for} \quad k = 1,2.$$

3. THE LINEAR OPINION POOL

Raiffa's discussion of the problem centres around use of the <u>linear</u> <u>opinion</u> <u>pools</u> or, in less technical terms, taking the arithmetical average of the experts' probabilities. Thus

$$\pi(.) = \tfrac{1}{2}p_1(.) + \tfrac{1}{2}p_2(.).$$

Here the experts' beliefs have been given equal weight, but, if DM thought it appropriate, he could use unequal weights:

$$\pi(.) = w_1p_1(.) + w_2p_2(.).$$

Simply adopting a linear opinion pool is not sufficient to prescribe a solution to the example, because the resolution provided to the conflict depends on whether the opinion pool is applied to the joint or marginal probabilities for A and B. Table 2(a) gives the result if the joint probabilities are averaged and then the marginal probabilities formed from these. Table 2(b) gives the result if the marginal probabilities are averaged and then the joint probabilities calculated under the assumption that A and B are independent. It can be seen that the results are distinctly different; and, in particular, averaging the joint probabilities does not maintain independence.

Raiffa, in the discussion, seems to favour the second course of action; namely, averaging marginal probabilities and preserving independence. But it should be remembered that his book is some sixteen years old and, since its publication, the literature on conflict of belief has grown considerably. He may now hold a different view.†

---

† In a letter Howard Raiffa has indicated that he accepts the general approach of section 6 of this paper.

	£ rises	£ falls	
	A	Ā	
Equipment:			
succeeds, B,	0.19	0.31	0.5
fails, B̄.	0.31	0.19	0.5
	0.5	0.5	

(a)

	£ rises	£ falls	
	A	A	
Equipment:			
succeeds, B,	0.25	0.25	0.5
fails, B̄.	0.25	0.25	0.5
	0.5	0.5	

(b)

Table 2   The DM's probabilities formed:

(a) by arithmetically averaging the experts' joint probabilities and then marginalising;

(b) by arithmetically averaging the experts' marginal probabilities and maintaining indepndence.

DM	£ rises A	£ falls $\bar{A}$	
Equipment			
succeeds, B,	0.25	0.25	0.5
fails, $\bar{B}$.	0.25	0.25	0.5
	0.5	0.5	

Table 3:   The DM's probabilities formed by geometrically
           averaging the experts' probabilities and normalising
           the result.

4.   THE LOGARITHMIC OPINION POOL

   There are other averages other than the arithmetical. For
instance, the geometric:

$$\pi(.) \quad \propto \sqrt{p_1(.)p_2(.)}$$

   Again we have given the experts' belief equal weighing, but
there is no a priori reason why we, or rather the DM, should do
this. The general logarithmic opinion pool allows unequal
weighting:

$$\pi(.) \quad \propto p_1^{w_1}(.) \; p_2^{w_2}(.)$$

where $w_1 + w_2 = 1$. Note that $\pi(.)$ is proportional to, not equal
to, the geometric average; $\pi(.)$ must be normalised so that it
has total mass of 1. In the example we take $w_1 = w_2 = \frac{1}{2}$.

   Again there is a choice to be made. $\pi(.)$ could be formed
by taking the geometric averages of the experts' joint
probabilities, the result normalised, and the marginal
probabilities formed. Alternatively, the geometric averages
of the experts' marginal probabilities could be formed, these
normalised and then the joint probabilities calculated under
the assumption that A and B are independent. In either case
the result in Table 3 is obtained.

   At first sight this would seem a promising approach to the
solution of Raiffa's example. Alas, that promise is not
fulfilled on closer examination. Genest and Wagner [8] have

just proved a theorem, which in rough terms states for the general case of n experts the following.

If $\pi(A) = f(p_1(A), p_2(A), \ldots, p_n(A)) \times$ normalisation constant,

where the normalisation constant may depend on the distributions $p_k(.)$, but may not depend on the event A, and if

independence preservation holds on a field of greater than 4 atomic events; then $\pi(.)$ is dictatorial.

It is worth pausing to be sure what this theorem is stating. For the result to hold, $\pi(.)$ is <u>dictatorial</u> if $\pi(.) \equiv p_k(.)$

for some k: in other words $\pi(.)$ depends on the opinion of only one expert: all others are ignored. This is clearly an unsatisfactory general solution to the problem of pooling opinions. $\pi(A)$ is restricted to a function of the numeric probabilities assigned to A by the experts. There is no other dependence on the event A. Were there to be any other dependence on A, then $\pi(A)$ would be a synthesis of the information on the likelihood of A provided by the experts and <u>some other information</u> on A. Since the logarithmic opinion pool is of the functional form considered in the theorem, it is clear that it provides a persuasive solution to Raiffa's example only because the field of events is so small.

What then is the way forward? One way might be to consider rather more general averaging operations to form $\pi(.)$. But we shall not follow that path. Rather I shall suggest reasons why I find any approach to the expert problem that relies on some averaging operator too simplistic to provide a reasonable solution.

5. DIFFICULTIES WITH GENERAL OPINION POOLS

We have mentioned two opinion pools: the linear,

$$\pi(.) = \sum_{k=1}^{n} w_k p_k(.), \qquad \sum_{k=1}^{n} w_k = 1,$$

and the logarithmic,

$$\pi(.) \propto \prod_{k=1}^{n} p_k^{w_k}(.).$$

(Usually in the logarithmic opinion pool $\sum_{k=1}^{n} w_k = 1$, but this is not necessary.) However, there are many other ways of "averaging" the probability distributions $\{p_k(.)\}$ into a single

probability distribution $\pi(.)$: in fact, an infinity of other
ways. All, to my mind, suffer similar faults, and these faults,
it should be noted, are not simply that opinion pools fail
certain 'desirable' properties such as independence
preservation. Almost without exception I do not find such
properties desirable. I shall argue against independence
preservation shortly; I have argued against other properties
elsewhere [6]. Rather my objection to opinion pooling is that
such procedures misinterpret the task facing the DM in the
expert problem.

So what problems do I see with opinion pools?

Firstly, all introduce some weights $\{w_k\}$ but do not
operationally define them. Usually in group choice problems
one can appeal to some principle of fairness, democracy or
a priori bargaining strength to deduce values for such unknown
weights. But here one cannot. In matters of preference one
may argue that all individuals should be treated equally. The
same arguments are much less persuasive in suggesting that all
forecasts, all beliefs about an uncertain future, should be
treated equally. In expertise, knowledge and forecasting
ability, all individuals are demonstrably not equal and it
would be foolish to pretend that they were.

How then are the $\{w_k\}$ to be determined? Clearly they may
be taken as reflecting the relative expertise of the experts,
but that really does not help in suggesting how they should
be measured. To say that one expert is 'twice as good' as
another is a figure of speech, not an arithmetic statement.
Several pragmatic solutions have been proposed, e.g. [22],
but to my knowledge none avoid a certain arbitrariness.
Moreover, the $\{w_k\}$ may reflect properties other than relative
expertise. For instance, two experts are seldom, if ever,
independent sources of information. They share a language and
hence common concepts; they probably have had similar
knowledge bases; etc. Should the $\{w_k\}$, whatever they are, be
adjusted to allow for this, to avoid an element of 'double
counting'?

Second, all opinion pools treat the experts' probabilities
as if they are honest, accurate reflections of the experts'
true opinions. But this is far from likely. The assumption
that the experts honestly and truthfully report their beliefs
is strong and essentially unverifiable. Moreover, once there
is more than one expert it is not clear that honesty can be
bought through a scoring rule [6]. Even if the experts do try
to honestly report their beliefs it is far from likely that

they will be able to do so accurately. For the DM to
interpret an expert's probability, the DM needs to know not
only what substantive knowledge the expert possesses but also
how well calibrated the expert is, i.e. how good he is at
encoding his beliefs as probabilities. The literature on
calibration suggests that few are good at this and that many
biases may creep into the assessment of subjective
probabilities and, one suspects, any verbal forecast [11].

The way around all these problems is, I believe, to
remember what De Finetti has said about subjective probability
[1]. Probability only has existence for the person whose
beliefs are modelled. Thus an expert's probability is a
probability for him, but a _piece of data_ for the DM. The task
facing the DM is to learn from the data that he gains, and in
the expert problem the data are the experts' beliefs.

6. A SOLUTION THROUGH BAYESIAN UPDATING

If the experts' probabilities are simply data for the DM,
then the DM should use these data to update his prior beliefs
about the events in a, and, and since we are implicitly
assuming his allegiance to the theory of subjective
probability, we may presume that he does this through Bayes'
Theorem. This has the immediate advantage, as we shall see,
that he can allow for the possibilities of miscalibration,
dishonesty and nonindependence of the experts through the
likelihood function. However, there is one further point that
we should remember. If the decision maker were one of the
experts we should allow for the possible correlation of his
beliefs with those of the others. Simply because we have
conceptually separated him from the group does not mean that
we should forget this possibility. His prior beliefs may be
correlated with those of the experts. [2,3,5,6].

The probability model that follows is developed informally,
mathematical details and methodological foundation are
presented elsewhere [2,3,4,5,6]. It should be emphasised at
the outset that the purpose of this particular model is to
suggest that independence preservation is not so clearly a
desirable property. There is no claim that the model is _the_
way in which to solve the example.

Applying Bayes' theorem to give the DM's posterior
probability for the joint event AB given the experts'
statements:

$\pi($AB$|$experts' statements, $H_o)$

$$\propto \ \pi(\text{experts' statements}|\text{AB},H_o) \times \pi(\text{AB}|H_o)$$

where $\pi(.)$ has been used as a generic symbol for any of the DM's beliefs and H_o is the DM's prior knowledge of all relevant issues. The DM's posterior probabilities for the joint events $A\bar{B}$, $\bar{A}B$ and $\bar{A}\bar{B}$ are evaluated similarly.

Given the experts believe that A and B are independent events, they really only give four pieces of information: $p_1(A)$, $p_1(B)$, $p_2(A)$ and $p_2(B)$. Moreover, we shall assume that a priori the DM's beliefs about A and B are independent:

$$\pi(AB|H_o) = \pi(A|H_o) \times \pi(B|H_o).$$

In fact we shall find it easiest to work in terms of log-odds:

$$\ell_1(A) = \ell n[\, p_1(A)/(1-p_1(A))\,]$$

with $\ell_1(B)$, $\ell_2(A)$ and $\ell_2(B)$ being defined similarly. We will also need the DM's prior log-odds:

$$\lambda(A) = \ell n[\, \pi(A|H_o)/(1-\pi(A|H_o))\,]$$

with $\lambda(B)$ being defined similarly. Finally we define two vectors: the first the DM's prior assessments

$$\underset{\sim}{\lambda} = \begin{pmatrix} \lambda(A) \\ \lambda(B) \end{pmatrix}$$

and the second the experts' log-odds:

$$\underset{\sim}{\ell} = \begin{pmatrix} \ell_1(A) \\ \ell_1(B) \\ \ell_2(A) \\ \ell_2(B) \end{pmatrix}$$

To update the DM's prior beliefs through Bayes' theorem we need to specify the four likelihood functions: $\pi(\underset{\sim}{\ell}|AB,H_o)$, $\pi(\underset{\sim}{\ell}|A\bar{B},H_o)$, $\pi(\underset{\sim}{\ell}|\bar{A}B,H_o)$ and $\pi(\underset{\sim}{\ell}|\bar{A}\bar{B},H_o)$. To produce these I suggest that we should enlarge the discussion from the assessment of the particular events A and B on the particular occasion that concerns the DM. Suppose that the DM imagines a sequence of occasions on which he and the experts give their subjective probabilities for events "similar to A and B". Thus he imagines a sequence of occasions on which he needs to predict the future value of the £ and the success of the

equipment. He may then consider the joint distribution of
his assessments λ and the experts' ℓ. Note that in this
mind-experiment the DM's prior assessments λ are random for
him because he lacks the focus of a particular occasion to
define them precisely. In [2,5] I discuss this mind-experiment
in greater detail. By considering such a sequence the DM may
produce four conditional distributions of belief, which we
shall assume are normal:

$$\begin{pmatrix} \lambda \\ \ell \end{pmatrix} \Bigg| \ AB \ \sim \ N(m_{AB},V)$$

$$\begin{pmatrix} \lambda \\ \ell \end{pmatrix} \Bigg| \ A\bar{B} \ \sim \ N(m_{A\bar{B}},V),$$

$$\begin{pmatrix} \lambda \\ \ell \end{pmatrix} \Bigg| \ \bar{A}B \ \sim \ N(m_{\bar{A}B},V)$$

$$\begin{pmatrix} \lambda \\ \ell \end{pmatrix} \Bigg| \ \bar{A}\bar{B} \ \sim \ N(m_{\bar{A}\bar{B}},V),$$

where $N(\mu,\Sigma)$ represents a normal distribution with mean μ and
covariance Σ.

To be clear what is meant by these distributions, consider
the distribution of $\begin{pmatrix} \lambda \\ \ell \end{pmatrix}$ |AB. This describes the variation the

DM would expect to see in his own prior assessments and those
of the experts in circumstances in which the joint event AB
subsequently occurs. Arguments suggesting why this
distribution might be near normal are given in [2,3,15]. The
mean vector m_{AB} used in the numerical example is given in
Table 4. Thus, for instance, the DM would expect Expert 1 to
give log-odds on A of 1.61 and log-odds on B of 0.20 in
circumstances in which A and B subsequently occur. Looking at
$m_{\bar{A}\bar{B}}$, we see that the DM would expect the same expert to give
log-odds of -1.61 on A and -0.20 on B in circumstances in
which neither A nor B subsequently occurs. If the variances
were the same, this would imply that expert 1 were better able
to discriminate circumstances that lead to the occurrence of A
than circumstances that lead to B. In fact the choice of

variances emphasises this discriminatory ability. Conversely,
Expert 2 is assumed to be better at forecasting B than A.

Mean log-odds of:	$m_{\sim AB}$	$m_{\sim A\bar{B}}$	$m_{\sim \bar{A}B}$	$m_{\sim \bar{A}\bar{B}}$
DM on A	0.41	0.41	-0.41	-0.41
DM on B	0.41	-0.41	0.41	-0.41
Expert 1 on A	1.61	1.61	-1.61	-1.61
Expert 1 on B	0.20	-0.20	0.20	-0.20
Expert 2 on A	0.20	0.20	-0.20	-0.20
Expert 2 on B	1.61	-1.61	1.61	-1.61

Table 4   The mean vectors $m_{\sim AB}$, $m_{\sim A\bar{B}}$, $m_{\sim \bar{A}B}$ and $m_{\sim \bar{A}\bar{B}}$ used in the
          example. Note that:

> log-odds of 0.41 correspond to odds of 6:4,
> log-odds of 1.61 correspond to odds of 5:1,
> log-odds of 0.20 correspond to odds of 55:45.

In the model the same covariance matrix has been assumed
for all four distributions. This like so much else in this
example is for mathematical convenience. More reasonable
assumptions are discussed in [14]. However, the form of V
used in the numerical example is important and should be noted:
see Table 5. Consider first the diagonal elements. These
reflect the precision expected in the assessments. Thus, note,
for instance, that expert 1 is more precise at forecasting
event A than event B. The within person covariances, i.e. the
off-diagonals within the dotted boxes, reflect the covariation
that arises between assessments because the same person is
making them. The other non-zero covariances reflect the
covariation that arises between assessments because, although
different people make them, they are on the same event.
Finally the zero covariances reflect that little covariation
would be expected between assessments made by different people
on distinctly different events.

The likelihood function $\pi(\ell \mid AB, H_o)$ etc. may now be
developed from the normal distribution (1). I have argued
in detail elsewhere [2,3,5] that the DM's actual prior log-
odds λ that he assesses on the occasion of his actual decision
are a suitable proxy for his prior knowledge H_o. Thus

$\pi(\ell | AB, H_o)$ etc. are formed by conditioning the normal distributions (1) on λ. The form of such conditional normal distributions are well known and may be found stated for this context in French [3]. Thus the version of Bayes' Theorem used to update the DM's beliefs becomes

$$\pi(AB | \ell, H_o) \propto \pi(\ell | AB, \lambda) \times \pi(AB | H_o)$$

with similar expressions holding for his posterior probabilities on $A\bar{B}$, $\bar{A}B$ and $\bar{A}\bar{B}$.

Row corresponds to
co-variance of log-
odds of:

DM on A	4.0	1.0	1.0	0.0	1.0	0.0
DM on B	1.0	4.0	0.0	1.0	0.0	1.0
Expert 1 on A	1.0	0.0	3.0	0.5	1.0	0.0
Expert 1 on B	0.0	1.0	0.5	6.0	0.0	1.0
Expert 2 on A	1.0	0.0	1.0	0.0	6.0	0.5
Expert 2 on B	0.0	1.0	0.0	1.0	0.5	3.0

Table 5   The covariance matrix V used in the example. The within person covariances are indicated by the dotted lines.

Tables 6 and 7 give, respectively a possible set of prior probabilities of the DM and the posterior probabilities that result when Bayesian updating is applied. Note that the events A and B are independent under the DM's prior beliefs and remember that the experts' beliefs give A and B to be independent. Yet the DM's posterior beliefs are such that A and B are no longer independent. Independence is not preserved. Moreover, note that this loss of independence is not some quirk of the numbers that I have chosen. Within broad limits, whatever λ and ℓ, providing that m_{AB}, $m_{A\bar{B}}$, $m_{\bar{A}B}$ and $m_{\bar{A}\bar{B}}$ are not all equal and providing that the non-zero covariances in V take some non-zero values, then the posterior beliefs will not preserve independence. Does this argue against the Bayesian formalism? I think not.

DM's prior	£ rises	£ falls	
	A	Ā	
Equipment:			
succeeds, B	0.3025	0.2475	0.55
fails, B̄	0.2475	0.2025	0.45
	0.55	0.45	

Table 6   The DM's prior beliefs in A and B as used in the example.

DM's	£ rises	£ falls	
posterior	A	Ā	
Equipment:			
succeeds, B.	0.702	0.179	0.881
fails, B̄.	0.098	0.021	0.119
	0.800	0.200	

Table 7   The DM's posterior beliefs after updating his prior beliefs (Table 6) with the information provided by the experts (Table 1). (That is λ and ℓ were calculated from Tables 6 and 1 respectively and fed into the normal model described in the text.)

First, remember that we are using probabilities to model beliefs. They are subjective probabilities, not objective ones. Thus probabilistic independence does not reflect some physical independence between events. Rather one event is probabilistically independent of another to the DM if knowledge of the occurrence or non-occurrence of the first would not affect his probability for the second. Now consider the problem facing the DM in Raiffa's example. He is concerned

with events A and B mainly, but he is also an observer of a
forecasting system: the two experts. Suppose that he learns
their beliefs and then learns that A has occurred. The
knowledge that A has occurred will enable the DM to evaluate
the quality of the experts' forecasts and that evaluation will
lead him to revise his opinion of their forecasting ability.
Given that he has revised his judgement of their forecasting
ability, it is natural, indeed inevitable, that he will see
their beliefs about B in a different light and, consequently,
revise his probability for B. In short, after hearing the
experts' opinions, knowledge of A will change his probability
for B: A and B are no longer independent.

Thus the Bayesian updating model suggests that a property,
independence preservation, which initially seems highly
desirable in any method of assimilating experts opinions into
the DM's own does not seem nearly so desirable on closer
examination. Had we time to examine other "desirable"
properties such as external Bayesianity or marginalisation
(see appendix), then we would have reached a similar conclusion.
The expert problem and its relatives are deceptively complex,
as indeed are all problems of resolving conflict.

Appendix:

A BRIEF GUIDE TO THE LITERATURE

In this paper we have concentrated on one particular
example and one particular issue, independence preservation.
The literature on conflict of belief is much wider and covers
many more issues. The purpose of this section is to provide a
brief guide to that literature. Much fuller surveys may be
found in [6,9,10,22].

Perhaps the first point to note is that we have been
discussing the expert problem, but there are other problems
in which the need to resolve conflict of belief may arise:
the group decision problem in which the group itself is
responsible for the decision, and the text-book problem in
which there is no focus of a decision and in which the group
is simply asked to report their beliefs in order to increase
public knowledge. In [6] I survey the literature on all these
problems and indicate that properties that are desirable in the
solution of one problem may not be desirable in the solution of
another.

In this paper we have concentrated on Independence
Preservation. Other discussions of this property may be found
in [8,12,13].

Two 'desirable' properties suggested in the literature are:

Marginalisation Suppose that interest is focused on a sub-σ-field of that over which the experts gave their beliefs. It seems intuitively appealing that the same synthesis of their beliefs be obtained whether (i) their beliefs are first combined into a distribution over the complete σ-field and then the marginal distribution taken, or (ii) the experts each give their marginal distributions over the sub-σ-field and a synthesis of these formed. Discussions may be found in [6,14, 17,21].

External Bayesianity Suppose that some objective evidence becomes available and that everyone concerned agrees on the likelihood function derived from this evidence. Then it seems reasonable to demand that the following two procedures give rise to some synthesis of belief. (i) Each expert updated his own belief through Bayes' Theorem and then the synthesis of these posterior distributions is formed. (ii) The synthesis of the experts' beliefs first is formed and then this updated through Bayes Theorem. This property is discussed in [6,7,14, 16].

In section 5 we mentioned the issue of calibration of the experts: how good are they at encoding their beliefs as probabilities. Some authors have suggested that exploring this issue carefully will lead to a simple solution to the expert problem. Others have disagreed [5,18,20] .

ACKNOWLEDGEMENT: I am grateful to Christian Genest and Carl Wagner for a pre-print of their paper.

REFERENCES

1. De Finetti, B., (1974) *Theory of Probability*. Vol. 1. John Wiley, Chichester.

2. French, S., (1980) Updating of belief in the light of someone else's opinion'. *J. Roy. Statist. Soc.* A143, 43-48.

3. French, S., (1981) 'Consensus of opinion', *Eur.J.Opl.Res.* **7**, 332-340.

4. French, S., (1982) 'On the axiomatisation of subjective probabilities'. *Theory and Decision*, **14**, 19-33.

5. French, S., (1986) 'Calibration, refinement and the expert problem', *Mgmt Sci.* (In press).

6.   French, S., (1985) 'Group consensus probability distributions: a critical survey' in Bernado, J.M., DeGroot, M.H., Lindley, D.V. and Smith, A.F.M., Eds. *Bayesian Statistics II,* North Holland, Amsterdam, 183-203.

7.   Genest, C., McConway, K.J. and Schervish, M.J., (1984) 'Characterisation of externally Bayesian pooling operators' Faculty of Mathematics, University of Waterloo, Canada.

8.   Genest, C. and Wagner, C.G., (1984) 'Further evidence against independence preservation in expert judgement synthesis' Department of Statistics, University of Waterloo, Canada.

9.   Genest, C. and Zidek, J.V., (1986) 'Combining probability distributions: a critique and annotated bibliography'. (with discussion) *Statistical Science,* (In press).

10. Hogarth, R.M., (1977) 'Methods for aggregating opinions' in H. Jungermann and G. DeZeeuw, Eds. *Decision Making and Change in Human Affairs.* D. Reidel Pub. Co., Dordrecht.

11. Kahneman, D., Slovic, P. and Tversky, A., Eds. (1982) *Judgement under Uncertainty: Heuristics and Biases,* Cambridge University Press, Cambridge.

12. Laddaga, R., (1977) 'Lehrer and the consensus proposal' *Synthese,* **36**, 473-477.

13. Lehrer, R. and Wagner, C.G., (1983) 'Probability amalgamation and the independence issue: a reply to Laddaga.' *Synthese,* **55**, 339-346.

14. Lindley, D.V., (1985) 'Reconciliation of discrete probability distributions' in Bernado, J.M., DeGroot, M.H., Lindley, D.V., and Smith, A.F.M., Eds., *Bayesian Statistics II.* North Holland, Amsterdam, 375-391.

15. Lindley, D.V., Tversky, A. and Brown, R.V., (1979) 'On the reconciliation of probability assessments (with discussion)' *J. Roy. Statist. Soc.,* A142, 146-180.

16. Madansky, A., (1978) 'Externally Bayesian groups' Unpublished manuscript, University of Chicago.

17. McConway, K.J., (1981) 'Marginalisation and linear opinion pools'. *J. Amer. Statist. Assoc.,* **76**, 410-419.

18. Morris, P.A., (1983) 'An axiomatic approach to expert resolution, *Mgmt. Sci.* **29**, 24-32.

19. Raiffa, H., (1968) *Decision Analysis,* Addison Wesley, Reading, Mass.

20. Schervish, M.J., (1983) 'Combining expert judgements' Technical Report No. 294, Department of Statistics, Carnegie-Mellon University.

21. Wagner, C.G., (1982) 'Allocation, Lehrer models and the consensus of probabilities' *Theory and Decision* **14**, 207-220.

22. Winkler, R.L., (1968) 'The consensus of subjective probability distributions' *Mgmt. Sci.,* B15, 61-75.

MODELLING CONFLICT WITH WEIBULL GAMES

M. Yolles

(Richardson Institute, Lancaster University)

1. INTRODUCTION

The Richardson arms race is a primitive example of a dynamic model of conflict. Its simplicity makes it unable to cope with representing the real world. A more complete approach is considered through the concepts of game theory, and a new stochastic methodology is introduced. In this measurements are obtained from Weibull frequency distributions, and the parameters and expected values become inputs to a dynamic system. Represented as a many-player game, it is sensitive to influences from the game environment.

In this paper we shall be concerned with formal (mathematical) modelling of conflict phenomena such as war. It may be said that in the social sciences a formal model is often made in isolation of others, thus contributing to the disjointed, uncohesive nature of formal theory. The consequence is that like weeds in a garden lawn, different models with few common parts or points of connection are scattered about history without having contributed to an integrated development. There are signs that this may be changing.

One of the surest ways of creating cohesive formal theory is by building models within the confines of a general systems methodology. This provides a superstructure in which structures can be built; thus, specific models can be created by introducing well-defined hypotheses to structure them.

In physics, new models of behaviour normally integrate in some way with accepted models, so that somewhere a mathematical connection can be demonstrated. If measurements are supportive of a new model then the latter may be accepted; if they are in contradiction to the old model, then a

theoretical rationalisation may occur.   Science is measurement
orientated.

Measurements of reality are taken within a broad framework
of commonly accepted ideas and methodologies, which may be
referred to as a measuring platform.   An ideal type of social
measuring platform is one with a statistical processing
mechanism.   Raw data are processed, and outputs are contextually
orientated pieces of information which can feed deterministic
models, that is those built through logically deduced or
heuristic propositions.   Another variety of model, based on
probability functions, is referred to as stochastic.   These
relate more immediately to data and can overcome the problem
that experimentalists complain about [25] :   that deterministic
models often tend to be too restrictive to adequately represent
reality.   More about this later.

One deterministic model which perhaps defined the start of
formal conflict theory was that of Richardson, who was
attempting to formalise the process of arms escalation.   He
proposed a simple growth model [21] representing an arms race
between two nations, akin to the population growth models then
being investigated.   In the Richardson model, change over time
in the arms utility of a player is determined by a linear
expression.   Since it involves a representation of changes with
respect to the utilities, it may be called a _kinematic_ model.
Solution of the model involves identifying the value of the
players' utilities for determinable values of movements in
arms levels.   Other than the fact that the model does not work
for modern arms races, its limitations lay in its assumptions
of kinematic linearity (see equation (1) below), and that an
ambiguity was introduced when developing a solution.

Equations such as Richardson's provide a primitive dynamic
extension to what was until quite recently a static
methodology: Game Theory.   Kinematic changes relate to dynamic
changes, where the effect of changes in a utility value over
time can cause movements in a related energy function.   Games
describing such movements can be referred to as dynamic.

Dynamic Game Theory is a normative methodology which tends
to offer optimal solutions to problems with an explicitly
defined structure and well-defined conditions.   Solutions to
game theoretic problems are normally the production of game
outcomes, and these may vary over time.   The game itself will
also have an outcome which for example in zero sum situations
takes the value zero;   such game outcomes are calculated
through some form of aggregation [2] , so that the value of a
game is calculated by aggregating value to individual players.
When the aggregation takes equilibrium values, then no net

exchanges occur between the game and its environment; game theoretical models usually adopt this as a condition. One very serious question concerns how an aggregation is decided. It may occur empirically, for instance through multiple regression or expected utility theory, or according to some deterministic decision; but however aggregation occurs, it normally is independent of the proposed model itself, and the hypotheses governing it are arbitrary - dependent upon the investigator. The most common guideline which is followed is that of the simplest aggregation (e.g. linear combination), and is adopted unless good reasons exist for one more complex. Another problem within game theory is that it currently has an inability to economically represent mutual outcomes for any finite number of players - an aspect of what is often referred to as the many-body or n-body problem.

Closely connected with this methodology is that of Differential Game Theory - which appeared in the 1960's. The two differ from each other in the way that problems are expressed, and in the techniques for obtaining optimal solutions. Aspects of differential games operate in the same way as do aspects of mathematical Control Theory, an approach developed independently elsewhere. Like dynamic game theory, this tends to be used in the context of idealised situations, where control parameters are evaluated on the basis of well-defined conditions. Modelling also looks towards situations of optimisation: a game model has a structure a part of which (the payoff, or in control theory the objective function, determined by the utility function) is either maximised or minimised over time. When a maximum or minimum has been achieved, the game has been optimised. In this sense an optimal solution is the 'best' solution. Distinctions between dynamic and differential game theories tend to be less clear as both develop and use common dynamic techniques of analysis. While both methodologies deal with games closed to exogenous influences, it has more recently been realised that systems modelled as games do not in reality frequently achieve closure, and so movements have been made towards systems theory consistent with open games [16] .

In this paper, a branch of differential game/control theory has been transformed into a new methodology able to simulate open games. It uses a modelling parameter for each player which determines the nature of the game, and which can be chosen according to hypothesis. Perhaps its most useful feature is that it permits a game to be played either normatively or stochastically by choosing a function to which the selections of data conform as a frequency distribution. Here, a stochastic game will be presented which is based upon the Weibull distribution. This provides a structure for the game, requiring

three parameters (one of which is a control) to be estimated
for each player. The controls can be selected in order to
simulate reality. This approach has been designed to
facilitate the modelling of open games for n players, each
playing one with the other through a player potential. This
enables an n-player system to be represented by n two-player
expressions. An attempt in this work will also be made to
assign meaning to the components of the system superstructure;
this should assist in the interpretation of symbolic explanations
and results.

2. MODELS AND GAMES

   Mathematical models are constructed from a set of
propositions expressed through a specific set of symbolic
relationships, and which operate through variables whose values
indicate the state or condition of the system that the model is
trying to explain. When two or more individuals (or groups)
interact over some object of attention and have outcomes that
are cojoint and uncertain, then the modelling situation may be
called a game and the individuals (or groups) its players. Let
a player be represented symbolically by R_i so that in a group
$R_0, i=1,2$ refers to the first and second players R_1 and R_2. The
state variable relating to R_i is x_i; in game-theoretic
approaches this is often referred to as a utility: that is,
the subjective value of a commodity or outcome for a particular
player. If a set of utilities can be assigned to those outcomes
that can accrue to the layer during the interaction, then the
ordered set is a utility function which defines part of a game
structure.

   Conflict models which have the appropriate ingredients may
be referred to as games. One very well known model of conflict
behaviour which has frequently been adopted as a component of a
game is that of the Richardson Arms Race, which attempts to
explain the way in which different nations can mutually
stimulate one another into the accumulation of armaments. For
a two-player system, the state variable for R_i is the arms
utility x_i where i=1,2. The Richardson expressions representing
an arms race may be called the kinematic equations since they
express the rate of change of R_i's arms utility, given by the
linear expression:

$$dx_i/dt = a_{ij}x_j - a_{ii}x_i + f_i, \quad i=1,2, \quad j=2,1 \tag{1}$$

By postulate, a_{ij} is R_i's defence coefficient, while a_{ii} represents the fatigue and f_i the grievance. The pair of linear equations for both players can be solved to give a bounded exponential growth to arms accumulation; they represent an arms race process which is under control when

$$a_{12}a_{21} < a_{11}a_{22}$$

A non-linear form of the Richardson equations has also been suggested, in which

$$dx_i/dt = a_{ij}x_j(1-b_{ij}(x_j - x_i)) - a_{ii}x_i + f_i \qquad (2)$$

where b_{ij} is the coefficient of submissiveness from R_i to R_j. Attempts have been made (in particular through the pages of J. Peace Science) to demonstrate the validity of these models, with limited success. Areas of weakness in the Richardson models have been expressed by a number of authors. For instance Arber [1] and Rappoport [20] have pointed out that the model is one of national behaviour which on its own does not explain how nations come to adopt their particular reactions.

According to Gillespie and Zinnes [12], the Richardson model can be expressed as a Nash cooperative game, a condition which can uniquely specify the type of arms race situation being modelled. Friedman [10] tells us that the essential features of many theoretical games include the conditions that (a) there are two or more players, (b) each wishes to maximise his own payoff, (c) each player is aware that the other player's individual actions can affect the payoff he receives, and (d) the interests of one player vis-a-vis the others are neither perfectly opposed nor perfectly coincident. Players satisfying (b) are said to be rational, and a game satisfying (c) can be represented across time as a dynamic system. There are different classes of game; in one of these, cooperative games, players have the capacity to make binding agreements.

For differential games, a structure is composed of the objective function, defined by a utility function over a period of time and the kinematic equations. This approach was first outlined in 1965 by Isaacs [15], and expressed in terms of capture and evasion (as in missile deployment) where players would win or lose a game. Isaacs' approach in optimising the objective function was to transform it using

Taylor's theorem, thus producing a form which directly
resembled that used within Hamiltonian-Jacobi optimisation
theory. It thus became clear that such an approach could be
implicitly included in this form of game, an idea which was
later developed by such authors as Friedman [9].

Hamiltonian-Jacobi methods are central not only to game
theory, but to aspects of control theory, independently
developed in the USSR. They contribute to the erection of a
mathematical superstructure within which a game structure is
defined, and involve its transformation into a Hamiltonian
function by the introduction of a set of auxiliary or slack
variables (also called multipliers) which assist in creating
optimal solutions. If an objective function is created by
postulate, then it becomes a relatively simple matter to
obtain a Hamiltonian function, often taken to represent the
energy in the system being modelled. For a Hamiltonian which
is time-invariant, certain relations with the kinematic
equations can be deduced - called the Hamiltonian canonical
system - which can make it relatively straightforward to
obtain optimal solutions if they exist. When the Hamiltonian
is implicitly invariant with time, the energy is said to be
conserved, and the dynamic system is conservative. Often, the
use of controls in the system enables optimality to be achieved
through game adjustment as well as through moves of strategy by
the players. To appreciate how the problem can be posed in
this theory, first hypothesise a utility function for a given
player and a set of kinematic equations like that of
Richardson's. For example, we may introduce two hypotheses
concerning two players R_1 and R_2. First, suppose that each
wishes to minimise the absolute difference between the arms
utilities $x_1(t)$ and $x_2(t)$ during the time t ranging between
(t',t''). That is, R_2 wishes to minimise

$$(x_1(t) - bx_2(t))^2 \qquad\qquad (3)$$

where b is some constant, which we can take to reflect the
proportion of arms R_2 requires to ensure national security.

Secondly, however, each player also wishes to minimise the
total amounts of both players, i.e. to minimise

$$(x_1(t) + x_2(t)). \qquad\qquad (3a)$$

Combining these hypotheses linearly gives the utility function

$$u = (x_1 - bx_2)^2 + c(x_1 + x_2) \qquad\qquad (4)$$

The next step is to consider the time period (t',t"), and define the objective function or payoff P' by the integral

$$P' = \int_{t'}^{t''} u(x_1(t),x_2(t)) \, dt \qquad (5)$$

subject to the Richardson equation (1) as constraints [14] on the problem. The object of the game is to minimise P'. It is useful to note from the definition of the utility function in this model that P' will change according to the value of x_1 which we may conceive as a control on R_2. Hamilton-Jacobi theory can be used to find optimal conditions in the game in the following way. Form a Hamiltonian 'energy' function

$$H(x_1,x_2,X) = u(x_1,x_2) + Xf(x_1,x_2) \qquad (6)$$

where X is a multiplier. To appreciate the meaning of X we should first appreciate the meaning of the word optimality. When payoff has been either maximised or minimised its value is optimal; the values of x_1 and x_2 which give this are optimal values; an optimal utility function $u(x_1,x_2)$ is one which makes the payoff optimal. Now, it can be shown [14] that X represents the sensitivity of an optimal utility function to changes in player utility value at a given point in time. Thus if P'_{opt} is an optimal payoff at a fixed point in time t, then $X = \partial P'_{opt}/\partial x(t)$. With respect to player R_2, optimal P' occurs when the dynamic expression can be satisfied

$$q \, \partial H/\partial x_2 = -dX/dt, \quad q \, \partial H/\partial X = dx_2/dt \qquad (7)$$

where q is simply an arbitrary control as suggested by Andronov [1]. If the utility x_1 of player R_1 is considered as a control on player R_2, then also

$$\partial H/\partial x_1 = 0. \qquad (8)$$

It may be noted that when (7) and (8) hold, then H of (6) becomes invariant in time. Since it represents an energy function, its invariance means that energy is conserved. An optimal payoff therefore occurs when the game is conservative.

To find an optimal solution, given a value for x_1 say, what
should an optimal choice of x_2 be? Sketching out a solution
for this, we can find $\partial H/\partial x_2$ by straightforward differentiation,
thus giving dX/dt in terms of X in a first order linear
equation which can be directly solved by standard methods. The
solution to the problem can now be found by differentiating H
with respect to x_1, putting the result to zero, and
substituting for X. Virtual simplicity itself!

The above example represents interaction between two players,
and can easily be extended to many (or N) players. Thus, given
the utility values of all but one (or N-1) players, an optimising
utility value can be found for the remaining player. The
optimal payoff can now be found. Different solution objectives
in the game, pursued through different manipulations or
additional propositions will permit other results to be
obtained. Examples of such modelling are provided by
Gillespie and Zinnes [12], and Brito and Intriligator [5,6,13]

This structured approach is immeasurably superior to the
simple kinematic model of Richardson. It is a type of
problem termed normative, where specific objectives are
identified explicitly and constitute an ideal rather than
realistic situation. Normative models create a whole
idealised symbolic world where players play according to
arbitrary universal rules often with similar perceptions and
over the same objective. The idealised nature of the
normative approach makes it generally unable adequately to
represent reality, for it is created in the abstract rather
than directly from the phenomenon being studied. One way of
remedying this is to replace the predefined deterministic
structure by a probability function structure, when the model
may be said to be stochastic. The distinction between
deterministic and stochastic models is well defined in general
terms [28]; in terms of this paper, normative models relating
to the Richardson model may virtually be considered as the
obverse of the stochastic model which will be developed here
[29]. In either type of model, generally, a structure is
hypothesised which may or may not be an adequate representation
of the phenomenon being modelled. Evaluation of the parameters
associated with the model is then made using statistical
techniques. From hypotheses these parameters will have a
contextual meaning, so that a phenomenon based interpretation of
the value it takes can be made. Thus, in equation (3) not only
will the meaning of b be required, but the significance of any
value it takes must be known.

For a many-player game, one can determine individual player properties rather than overall game effects. Establishing overall game effects then requires some arbitrary form of aggregation of utilities. Mostly, aggregations are created according to some arbitrary static linear averaging formula of the state variables. For example, one might say that the aggregate utility value $x_o(t)$ for two player utility values $x_1(t)$ and $x_2(t)$ is given by

$$x_o(t) = p_1 x_1(t) + p_2 x_2(t) \qquad (9)$$

where if $p_1 = p_2 = 1/2$ then we have a simple average, or if p_1 and p_2 are probabilities then we have $x_o(t)$ as an expected value. Some thought has been given to more satisfactory general forms of non-linear aggregations which occur at the dynamic level [2], an approach which this paper supports and develops.

If a game is a representation of a reality which can be perturbed away from an initial normative model then any initial optimisation will be lost. A game able to represent such perturbation is essentially an open game. In the case where Hamiltonian energy is the only representation of whether a game is open or closed, then optimality is an indicator of closure.

Each of these problems shall be considered in due course.

3. EXTENDING GAME CAPABILITIES

In this section, we shall explain how the above theory can be extended in order to overcome some of the problems highlighted above. It is appropriate to begin this by initially rejecting the idea of proposing an arbitrary utility function. In fact we can go further and say that the utility function cannot be determined in the abstract. This has immediate consequences for the meaning of the multiplier introduced in equation (6). Earlier the multiplier $X = X(t)$ at the given point in time t was said to be the sensitivity of an optimal utility function to changes in player utility. To provide X with a new meaning, suppose that we replace the optimal utility function by a universal player potential P. We could now perceive that the multiplier is the sensitivity of the potential to changes in utility value. In fact however it can be argued that because of the nature of P the meaning of X rather changes to the obverse of the normative interpretation: the potential P responds to an initial change x(t) by making an initial change X(t). Now suppose that there are N players R_1, $R_2, \ldots R_N$, each represented by R_i for i=1, N and partitions R, of the set of possible players. Suppose also that P can be

partitioned into a set of local player potentials P_i with
corresponding multipliers X_i, and each P_i corresponds to
player R_i in such a way that R_i and P_i are interactive pairs.
Thus all interactions between R_i and the other players are
represented by the interaction between R_i and P_i. It
transpires that now the relationship between each player and
that player's utility value is the same as that between the
local player potential and the associated multiplier. That is,
if x_i is R_i's utility move in the game, then X_i is P_i's
corresponding change. The conservative Hamiltonian condition
which generalises (7) still holds for this amendment, and can
be written in the form

$$q\ \partial H/\partial X_i = dx_i/dt\ ,\quad q\ \partial H/\partial x_i = -dX_i/dt,\ i=1,N \qquad (10)$$

It may be noted that since P is not fixed as in normative
theory, adjustments can be made to the value of X with
appropriate adjustments being implied in P.

Consider now that the game becomes disturbed away from the
conservative condition. For a fixed potential P it may be
possible to adjust for the perturbation, thus ensuring that
the Hamiltonian canonical system is satisfied. To do this
let us suppose that when the game becomes perturbed q can be
replaced by q_i and q_i' for the perturbed system so that

$$q_i\ \partial H/\partial X_i = dx_i/dt,\ q_i'\ \partial H/\partial x_i = -dX_i/dt,\ i=1,N \qquad (11)$$

Clearly the game becomes conservative when

$$q_i = q_i' = q.$$

Our interest lies in transforming (11) into a stochastic
model. Firstly, however, we wish to define two further
requirements: (i) to find a way of operating (11) without
predefining an objective function as in (5); (ii) to establish
a way of aggregating parameters and variables in a game
dynamically.

To satisfy (i) we can rewrite the left hand side of (11) as

$$w_r m_r\ \partial H/\partial x_r = dx_r/dt,\ r=0,N \qquad (12)$$

where

$$m_r = q_r (\partial H/\partial X_r)/(\partial H/\partial x_r). \qquad (13)$$

In fact we may note that this definition of the modelling parameter makes it an index of interaction between the local player potential P_i and the paired player R_i. The rate of change of each player's move is given by

$$dx_r/dt = w_r x_r/m_r \ , \ r = O,N. \qquad (14)$$

The form of the modelling parameter in (13) is not arbitrary, for it (a) allows us to regain the Hamiltonian canonical form, and (b) permits modelling to occur at the dynamic rather than state level which can be an advantage when modelling relativistically, and which will in due course provide an entry into stochastic modelling. It may be said that with these formulations the shape of the potential P_i is such that $\partial H/\partial x_i$ can be expressed in terms of a sum of squares of player indexes of interaction.

To satisfy (ii) above, we may define dynamic aggregation for a subset of players of size n (not greater than the total set of players N) by

$$(\partial H/\partial x_o) dx_o/dt = \sum_{i=1}^{n} (\partial H/\partial x_i) dx_i/dt \qquad (15)$$

which implicitly defines dx_o/dt, or by (14) x_o. It may be noted that this expression can be written in terms of the modelling parameters by using (11). From here on, for simplicity we shall select n=N, so that our set of aggregated players represents the whole set of players in the game.

4. ENERGY AND ENTROPY IN CLOSED AND OPEN GAMES

Consider that the Hamiltonian represents all of the energy within a game: that is (a) the resources, (b) the structures which have been built and require upkeep, and which are a precursor for action, and (c) the kinematic energy of action itself. The energy component may not be the only descriptor of a system. A more general way of describing the condition of a system is through entropy; defined in terms of probabilities, this is a distribution of events within a game, and is taken as a measure of disorder or uncertainty. Like

Hamiltonian energy, entropy can be thought of as a differentiable function; when its rate of change dU/dt=0, then there is no net flow of entropy over time and the value of entropy in the game is constant. When this occurs the game is said to be closed (or net closed), for then no net exchanges occur between the game and its environment. Entropy is therefore an important concept, and further reading on this subject is recommended ([19],[8],[26],[16]).

Entropy has been used as an indicator of condition in the social sciences before now. For instance Coleman [7] used it in his study on voting behaviour as a measure of political uncertainty, and it has also been used by Galtung [11] in a more theoretical context.

Entropy is a descriptor of only one part of a system. A complementary descriptor is that of information I. The nature of I was conceived within information theory as devised by Shannon [22]; here I represented a measure of the homogeneity of a system, as opposed to the heterogeneity relating to uncertainty or disorder and thus associated with entropy [17]. The relationship between U and I has been considered by a number of people including Brillouin [4] who called the latter negentropy.

Axiomatically, one can propose that energy is consequential not only to behaviour but also to information. Now information may be regarded as a constraint on entropy, and using statistical mechanics it is possible to identify a relationship between both of these and energy. Let us propose that the Hamiltonian H represents the total energy in a game and that it has a rate of change dH/dt. Then the relationship between the entropy (U), energy (H) and information (I) can be given in terms of their rates of change by

$$dU/dt = vdH/dt - dI/dt \qquad (16)$$

for some game parameter v. It should be noted that this is a general system equation; when applied in physics 1/v is called temperature, and takes values from a standardised measuring platform. In a similar way in conflict situtations, 1/v could be conceived as the 'temperature' of a conflict. How values for such a measure might be arrived at directly is not clear; however, in a system where other variables/parameters are known, the 'temperature' could be determined consequentially. This equation can eventually be arrived at using an approach such as described by Socrates [24] among others, proposing that information constrains the distribution of events as if it were a directed field. When the left hand side of (16) is zero,

then most generally there is a steady state between the
information process and energy flow. On the other hand a zero
information process will mean that entropy and energy rates
of change move together macroscopically: here, a game which
is conservative in the Hamiltonian sense will have a zero
entropy rate, and since this implies no net exchanges with the
environment, the game is closed. An energy-conservative
game is not closed if an information flow occurs into it from
its environment. It can be shown that a possible way of
practically testing whether or not a conservative game is
closed (with no net exchanges) is to determine if the ratio of
the flows of energy and information for each player are the
same.

6. WEIBULL GAMES

Conditions can be arranged such that the index of
interaction for each player can be expressed in terms of
information flows. This gives the clue as to how these
indexes can be evaluated. If p_i is a probability representing
the homogeneity of events relating to player R_i, then the
information associated with the player is given by

$$I_i = -Ln \ p_i. \tag{17}$$

If p_i is an analytical function then one may quite easily
model the system by analytical hypothesis. For instance, we
may propose that p_i represents a Weibull probability,
effectively involving two parameters and an expected value.
This choice has been used previously in conflict analysis
with some success [18] . Of the two parameters, one (h)
relates to organisational process, and the other (a) to
structure. The probability distribution is not dissimilar to
that of Poisson, and is given by the natural logarithm of p
as

$$Ln \ p = -hx^a \tag{18}$$

for the vectors p,h,x and a, each having components $p_i, h_i, x_i,$
a_i associated with player R_i for i=1,n. The meanings of h and
a will be considered in detail later. This distribution is
normally fitted to a data set, and estimates for the two
parameters and the expected utility x made.

The Weibull information flow is given by

$$\partial I/\partial x = ahx^{a-1}. \tag{19}$$

Since (19) is a vector expression, equations like this
exist for each player R_i. These expressions may be substituted
into the modelling parameters and an appropriate q may be
chosen for the game. It should be noted that in the Weibull
game q has been selected to simplify the system such that

$$1/m_i = (\partial I/\partial x_i) \sum_{j=1}^{n} x_j^{a_j + c_j} \tag{20}$$

where for each j, c_j is the control parameter which has been
introduced through q, the values of which may be chosen such
that for some c, $c_j = c$ for all j. From this definition of
m_i we can now determine expressions for dx_i/dt and $\partial H/\partial x_i$. By
analysing the system it is also possible to come up with
information on when a conservative game has occurred. In fact,
this happens when

$$w_i x_i \partial I/\partial x_i = w_{i+1} x_{i+1} \partial I/\partial x_{i+1} \tag{21}$$

It is clear that for an open game, if an appropriate
adjustment to the trajectory through w_i is possible, a closed
game will develop. Defining the weighting w such that when
all the w_i are equal then $w = w_i$ for all values of i, then w may
be chosen arbitrarily as a game adjustment or normalisation.

7. ATTRIBUTING MEANING

It is necessary for the components of any modelling system
to have meaning assigned, so that an interpretation can be
made of what is going on. This is so for Weibull games.

The Weibull distribution has two effective parameters, one
of process h and one of structure a. The structural parameter
may take positive values and can be seen as follows:

0 < a < 1 represents a negentropy decrease

a = 1 represents a negentropy constant

1 < a < ∞ represents a negentropy increase

In terms of modelling, the meaning of such values for a will
depend upon the meaning of the dynamic system itself. For
instance, if we are investigating a conflict situation between

n players, each having utilities with Weibull probability, then for a player object of peaceful settlement, a decrease in negentropy is consistent with a reinforcement of peace, while an increase is consistent with the likelihood of violence breaking out.

The variables of a game may be classed as having either active or passive properties. The former relate to explicit events or actions while the latter relate to states of internal events, for instance by analogy with the emotion in the mind of an individual. Meanings will now be postulated for the list of variables so far introduced, though they will require adequate practical demonstration. The validity of these meanings can be further considered in the light of a unit system [27] .

1) Active.

x_i : Utility for R_i; a subjective value of a commodity or outcome. It can also be thought of as R_i's utility choice in the game.

I   : The information or negentropy of the game, determined by the homogeneity of events. Its change over time implies change in the level of information in the game.

$\partial I/\partial x_i$: This is the information flow for player R_i. Its change with x_i is the information flow with respect to player R_i, and may represent the manoeuvrability of R_i in the game. Petersen [18] would refer to it as the intensity of the process.

U   : The uncertainty or disorder or entropy in the game, representing a measure of the heterogeneity of events. Its rate of change will be indicative of the degree of closure of the game.

$\partial U/\partial x_i$: R_i's entropy flow, which could represent a change in R_i's organisational condition, and implies a structural disturbance.

2) Passive.

dx_i/dt : The rate of change of R_i's utility. How
swiftly such change occurs may be connected
to the ability or freedom to make decisions;
it will be seen that the absolute value of its
inverse could represent the freedom of
decision in conflictual contexts.

H : The Hamiltonian energy of the game. This
may be equated with system energy and related
to power.

$\partial H/\partial x_i$: This is a directed change in game energy for
R_i, and can thus be equated to player reaction
between a single player and the remaining
players within the game.

e_i : Emotiveness of R_i, which can be measured by
the absolute value of

$$\{(\partial H/\partial x_o) - (\partial H/\partial x_i)\}/(\partial H/\partial x_o) + (\partial H/\partial x_i)\}$$

It represents the player reaction away from a game
norm or aggregate.

The parameters of the game may be identified as follows:

a_i: Weibull structural parameter.
In a conflictual context its value can by hypothesised
to relate to policy position for R_i. It could be
measured say on a scale of hawkish/dovish policy.

h_i: state of interactive process for player R_i. Under
certain circumstances, for instance by comparing
the Weibull kinematic equation to the Richardson
model, $w_i h_i a_i$ can be conceived as a perception of
threat for R_i.

c_i: game direction connected with the structure.
In fact it magnifies the reaction of players
R_i in interaction. When $c_i=c$ for all $i=1,n$,
then c acts as a filtering magnifier, blowing
up larger values of reaction more than smaller
ones.

We have considered the case of closed games, when dU/dt is
zero, and it is worth mentioning the possibility of one
interpretation of this. Frequently when parties to a dispute
are locked into a conflict, the dynamics become a consequence
of inertial forces independent of any potential flows from the
game environment; they thus become predictable. Lock-in
conflicts [23] may thus be implied when net game closure
occurs.

9.   MEASURING A SYSTEM

One way in which measurement of an interactive system can
be made is through the use of a probability distribution.
Here we use the Weibull distribution in a measuring platform,
and the reader is directed to a work by Petersen [18] for
more information on this. In order to describe how
measurements can be taken and processed dynamically to show
the consequence of player interaction, we shall refer to a case
study by Petersen [18] . The case study concerns the
conflict between China and India which started during the late
1950's and ended in the early 1960's, with a very brief
war in 1962. It concerned a dispute over a region of boarder
territory which was annexed from China by Britain during the
days of her empire. During the conflict a large number of
diplomatic communiques were issued by both sides and in the
mid 1960's India published them as a collection to validate
her position. The set of data was complete, and hence an
attractive proposition to investigators. A number of
significant studies were undertaken using the data, and during
this time it was demonstrated that the time delay between the
issue of diplomatic communiques provided a good indicator of
the conflict process going on. Petersen took the data for one
player (China) and tested it to see how many time-adjacent
Weibull distributions could be established within it. Various
tests were set up to ensure a goodness of fit each time. Each
distribution which resulted was called a phase. The same was
repeated for the data of the other player (India). Petersen
used the expected utility values as eigen-values representing
the state of player moves, and the parameters and other
outputs of the model to explain player moves. This was
validated with historical narrative. The phase time bounds
did not match, but this was not a particular problem.

In another study [27] , Petersen's data were used to test
the Weibull game model rather than to investigate the conflict
itself. First, phases were broken down for each player to have
the same time bounds; thus more phases were created than
originally obtained, but with some duplication. This permitted
a one-to-one phase/time arrangement, essential if the
dynamics of the two players were to be investigated across

time, and interactive analysis undertaken.

The next problem was to determine values for the weightings
on the kinematic equations, and these were assumed to take the
value of one throughout the dispute. This is not a problematic
assumption in this case because the weighting is arbitrary,
being an additional degree of freedom brought into the
methodology to increase modelling flexibility. Evaluation of
the control parameters was a further requirement, which in this
case was obtained by comparing reaction outputs to a quantified
historical narrative. Choosing values for the controls for
each player is really a matter of hypothesising the 'intent'
of each player.

10.   CONCLUSION

This work was conceived as a method for simulating real
complex conflict situations, and established in such a way as
to permit individual player modelling in a game. In the
normative method upon which our approach is based, an
objective function is proposed, and to solve the system
assuming it to be closed, an auxiliary variable is introduced.
Here, we propose that in general the system is not closed,
so that the system can be influenced by exogenous or game
external phenomena. By proposing that the objective function
cannot be determined, and is thus replaced by a player
potential, it is found that the auxiliary variable takes the
role of a move by a player potential, and each player plays
with the player potential rather than any other individual
player. An aggregation facility has also been introduced,
thus in principle permitting complex systems to be viewed
piecemeal. Finally, the use of an entropy function has
enabled stochastic games to emerge, permitting the introduction
of the Weibull game.

In normative games, Hamilton-Jacobi theory is not very much
different from open game theory, in that the latter has at
least the same number of degrees of freedom as the former in
selecting structural elements of the game to adequately
describe it. Since the latter develops from the former, this
is not surprising. However, a feature of open game theory is
that aggregations appear as a consequence of the superstructure,
so that subgroups of players in a game can (in principle at
least) be described as playing with other subgroups within
the game. Thus, in a game of four players say, it should be
possible to postulate a coalition of two playing with a third
player, and the resulting coalition playing with the
remaining one. Alternatively, given an n player game
simulating reality, it should be possible to determine the
feasibility of coalitions forming between a player subset by
investigating stability conditions.

Perhaps as important a feature of open game theory is that it permits stochastic games to emerge. Here, not only is the implicit structure of the game determined non-symbolically, but it is quite possible for an investigator to be non-mathematical in order to simulate reality with a dynamic model. The very selection of data itself presupposes an object of attention which implies a structure, and a meaning for an optimal solution; given a computer package with adequate checks and safeguards, an investigator needs only to be sure of his data, to be able to determine controls, and to be able to interpret numerical results.

The Weibull distribution is perhaps one of the more flexible available, though it will not always be demonstrably satisfactory. However, set up within the modelling parameter the Weibull game becomes a dynamic tool to explain reality from empirical observations.

The game operates by separating the set of events into appropriate time components or phases. This is a consequence of the way in which Weibull measurements are taken, and would seem to be suitable for the investigation of social interactions. While the dynamic system can be used generally for the simulation and consequential explanation of real phenomena, it may well have more value in the creation of scenarios for the anticipation of possible developments and their consequences. In making scenarios of conflictual situations it is essential to know the meaning of different values of the Weibull parameters. Selection of their values can then be made heuristically, or by proposing some time variation treatment.

It may be seen that compared to the elementary Richardson arms race models, the theory here is complex because it tries to model real complex behaviour. This is the more so because it attempts to look at many players interacting in an open system. While few methods exist which try to apply such ideas to reality in the social sciences, it is this which makes the methodology flexible. There is no panacea for social modelling, however. All modelling approaches suffer from some difficulties and this one is no exception. Two problems exist here. The first relates to the setting up of the Weibull distribution within a statistical measuring platform under a number of suppositions. Various tests can then be instigated to try to ensure that the suppositions are not violated. Some authors question the degree to which such tests are adequate. The second problem lies within the dynamic system itself, once suitable measurements have been estimated. The direction of the game will determine what reaction and other dynamic output values are generated, and this is dependent upon the control values. While their estimate is essential, the methods for so doing are not sufficiently precise as yet.

FOOTNOTES

1. A fundamental distinction between deterministic and stochastic models lies in their causalogical definitions [Chacko, G.E., 1971, Applied Statistics in Decision Making, Elsevier]. In the former, given knowledge of the initial state of a system, then assuming isolation its future can be deduced; in the latter, any one of a set of futures can be deduced since each is probabilistically identified. In particular, if a deterministic model describes states of a system which are time dependent, as say $z(t)$, then the stochastic model will describe states which are also dependent upon environmental outcomes ζ as say $z(t, \zeta)$, where ζ is a random variable.

2. In the work here a further distinction exists between the types of deterministic and stochastic models considered. The Richardson model is just a "local" arms race application of a general simple constrained exponential growth model, formulated according to hypotheses. It is thus deterministic with parametric evaluation coming from a best fit to a set of data. In the Weibull stochastic model, a general approach is adopted, for which under best fit guidance the parameters of the probability function can be evaluated. The generality of the Weibull distribution means that a reasonable fit to the data is likely. The current difficulty lies in assigning contextual meaning to the Weibull parameters within specific areas of study. In this situation the Richardson model provides an obverse aspect to that of the Weibull game: in the former we have meaning and seek appropriateness, while in the latter under best fit we have appropriateness and seek meaning.

REFERENCES

[1]   Arber, Jr. H.R., (1968) The Structure of Social Action in an Arms Race. Conference paper at North American Research Soc. (International), Cambridge, Mass., USA.

[2]   Aoki, M., (1976) Optimal Control and System Theory in Dynamic Economic Analysis. North Holland Pub. Co., NY.

[3]   Andronov, A.A., (1973) Quantitative Theory of Second Order Dynamic Systems. Israel Programme for Scientific Translation.

[4]   Brillouin, L., (1967) Science and Information Theory.
Academic Press Inc., NY.

[5]   Brito, D.L., (1977) Nuclear Proliferation and the
Armaments Race. *J. Peace Science,* **2**,2,231-38.

[6]   Brito, D.L. and Intriligator, M.D., (1974) Uncertainty
and the Stability of the Arms Race. *Annals Econ. and Soc.
Measurement,* **3**,1,277-92.

[7]   Coleman, S., (1975) Measurement and Analysis of
Political Systems: a Science of Social Behaviour. Wiley, NY.

[8]   Davies, P.C.W., (1975) Cosmological Aspects of
Irreversibility. In Entropy and Information in Science
Philosophy; ed. Kuba, T.L., Zeman, J. Elsevier.

[9]   Friedman, A., (1971) Differential Games. Wiley.

[10]   Friedman, J., (1983) Oligopoly Theory. Cambridge Univ.
Press.

[11]   Galtung, J., (1975) Peace: Research, Education, Action.
A 5 volume set. International Peace Research Institute,
Oslo.

[12]   Gillespie, J., Zinnes, D., (1975) Progress in
Mathematical Models of International Conflict. Synthese,
31, 289-321.

[13]   Intriligator, M.D., (1964) Some Simple Models of Arms
Races. General Systems, 9,143-49. 1971, Mathematical
Optimisation and Ecnomic Theory. Prentice Hall. 1975,
Strategic Considerations in the Richardson Model of Arms
Races, J. Pol. Econ., 83, 339-53. 1975, Applications of
Optimal Control Theory in Economics, Synthese, 31,271-88.

[14]   Intriligator, M.D., (1971), Mathematical Optimisation
and Economic Theory. Prentice-Hall Inc.

[15]   Isaacs, R., (1965) Differential Games, Wiley.

[16]   Nicolis, G., Prigogne, I., (1977), Self Organisation in
Non-equilibrium Systems, Wiley.

[17]   Peters, J., (1975) Entropy and Information Conformities
and Controversies. In Entropy and Information in Science
and Philosophy; ed. Kuba, T.L., Zeman, J.; Elsevier.

[18] Petersen, I.D., (1981) The Dynamic laws of International Political Systems, 1823-1973, Inst. of Political Studies, University of Copenhagen, Copenhagen.

[19] Prigogne, I., (1962) Thermodynamics of Irrevesible Processes. Wiley.

[20] Rappoport, A., (1957), Lewis Fry Richardson's Mathematical Theory of War. J. Conflict Resolution.

[21] Richardson, L.F., (1961) Arms and Insecurity, Boxwood, USA.

[22] Shannon, C.E., (1948) A Mathematical Theory of Communication. Bell Syst. Techn. Journal, 27.

[23] Smoker, P., (1966) Time Series Analysis of Sino-Indian Relations. J. Conflict Resolution.

[24] Socretes, G., 1971, Thermodynamics and Statistical Mechanics. Butterworths, London.

[25] Teditchi, J.T., et al., (1973) Conflict, Power, and games; the experimental study of interpersonal relations. Aldine pub.

[26] Ugemov, A.I., (1975) Problems of Directions of Time and Laws of Systems Development. In Entropy and Information in Science and Philosophy; ed. Kuba, T.L., Zeman, J.; Elsevier.

[27] Yolles, M.I., (1985) Simulating Conflict with Weibull Games. In the 1984 conference proceedings of IMACS European Simulation Meeting on Simulation in Research and Development, Eger, Hungary. Ed. Javor, A., North-Holland Press.

2. ARMS RACES AND INTERNATIONAL SYSTEMS

Seagoon: "But Sir, back in England they told me all was well"

Bloodnok: "Back in England, all is well. It's here where the trouble lies."

(More Goon Show Scripts, Spike Milligan, Woburn Press 1973)

RICHARDSON REVISITED

Ian Sutherland

(Medical Research Council Biostatistics Unit, Cambridge)

INTRODUCTION

The efforts of most pioneers tend to be distorted as time
passes, or superseded by later work, or even forgotten. I
do not think that any of these fates has befallen Lewis
Richardson, but it is certain that his work has never been
well known nor, as far as I am aware, have any major advances
been made beyond the point he reached. My excuse for this
contribution is that it is worthwhile to remind ourselves
what Richardson achieved, and to consider whether his approach
is still relevant and whether it can be developed in the
light of world events during the thirty years since he died.

I have also a particular personal reason for wanting to
review Richardson's work. My first acquaintance with it was
in 1940 through his newly-published monograph, Generalised
Foreign Politics, which was later amplified into the microfilm
and book, Arms and Insecurity. I was a mathematics
undergraduate, a Quaker, and a pacifist, soon to register as
a conscientious objector, in a country at war. Not
unnaturally in these circumstances, the monograph made a deep
impression on me. The mathematical modelling was convincing
and the agreements with observation were disturbing. The
war appeared (almost) as an inevitable consequence of the
failure of politicians to appreciate the elegance of
Richardson's differential equations, to understand their
implications and learn from these, and to take appropriate
action. What was so depressing was of course that there had
been no opportunity for the politicians to do any of these
things, even if they had wanted to, since the monograph had
not appeared until war was already inevitable. My reason
for returning to Richardson now, towards the end of a working
life as a statistician in a field quite different from politics

and international relations, is simply to judge, if I can,
whether his ideas still have the validity they seemed to me
to have in 1940, and if so, whether any advances are possible
from the position he had reached by 1953.

I confine myself to Richardson's mathematical studies of
the psychology of international relations, as exemplified
particularly in arms races. He also made extensive and
revealing descriptive analyses of the magnitude and nature
of wars, which are included in his book Statistics of Deadly
Quarrels, but these are less relevant to the present theme.

THE BASIC EQUATIONS

Richardson started by considering the interactions of two
nations.

"The various motives which lead a nation in time of peace
to increase or decrease its preparations for war may be
classified according to the manner of their dependence on
its own existing preparations and on those of other nations.
For simplicity the nations are here regarded as forming two
groups. There are motives such as revenge or dissatisfaction
with the results of treaties; these motives are independent
of existing armaments. Then there is the very strong motive
of fear which moves each group to increase its armaments
because of the existence of those of the opposing group.
Also there is rivalry which, more than fear, attends to the
difference between the armaments of the two groups rather
than to the magnitude of those of the other group. Lastly,
there is always a tendency for each group to reduce its
armaments in order to economise expenditure and effort.
What the result of all these motives may be is not at all
evident when they are stated in words. It is here that
mathematics can give powerful aid."

(Richardson, 1939)

This leads him directly to a pair of differential equations
(originally published in 1935) to express changes with time
in the level of armaments, or the 'defences', x and y, of the
two nations, namely:

$$dx/dt = ky - \alpha x + g$$

$$(1)$$

$$dy/dt = \ell x - \beta y + h$$

The first of these equations summarises the contention
that the rate of increase of the 'defences' of the first
nation, dx/dt, is proportional (with a positive 'defence
coefficient' k) to the size of the 'defences' of the second
nation, y, which are interpreted by the first nation as
'threats'; at the same time the rate of increase is restrained

by the 'fatigue and expense' to the first nation of maintaining
its own defences, which is proportional to x, with a positive
coefficient α; finally the rate may be influenced by some
grievance of the first nation against the second, independent
of x and y, represented by the constant g, which may be
positive or negative. The second equation summarises the
reciprocal contention on the part of the second nation,
with coefficients ℓ, β, and a constant h. These are plausible
contentions - what are the implications of the equations?

There are certain special cases worth mentioning first.
An absence of armaments (x = y = 0) and grievances (g = h = 0)
implies that x, y remain zero - permanent peace resulting from
disarmament and satisfaction. However, a state of mutual
disarmament without satisfaction (only x = y = 0) is not
permanent. Nor is the state of unilateral disarmament (y = 0),
for y will not remain zero if h is positive.

Before proceeding to the general case, Richardson pointed
out that, in the equations (1), x and y were not necessarily
positive, and considered whether any meaning could be attached
to 'negative preparedness for war'. He suggested that this was
not 'peace, in the sense of a mere tranquil inattention to the
doings of foreigners', but implied that the nation 'directs
towards foreigners an activity designed to please rather than
to annoy them', namely 'co-operation'. The variables x, y
should therefore be thought of not simply as 'defences' or
'threats' (depending on the viewpoint) but as 'threats minus
co-operation', which might be measured as a balance between
armaments and trade.

Richardson examined the general case in terms of the locus
of x, y in a plane, for various values of the coefficients.
The equations define two linear 'barriers', intersecting at a
'point of balance'. The point (x,y) is confined to one of
the sectors defined by these barriers, and either drifts
away from the point of balance, implying an arms race
('instability') or towards it, implying a balance of power
('stability'). Figure 1 illustrates four unstable situations.
The shaded areas are those in which dx/dt or dy/dt, or both,
are positive, and the arrows represent the direction of
movement of the point (x,y). In each of these four situations,
with k = ℓ = 2, α = β = 1, and various values of g and h, the
point, once in the cross-shaded area, moves away from the point
of balance, that is, x and y both increase, and the nations
are involved in an arms race.

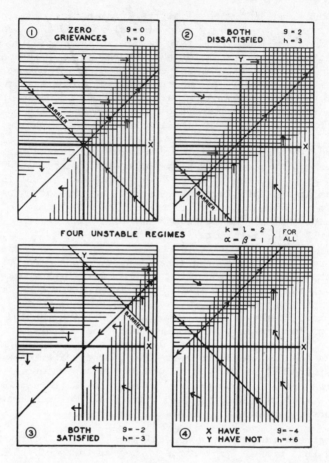

Fig. 1 Four examples of unstable configurations with the
armaments expenditures of the two nations moving
away from an unstable point of balance (from
Richardson, 1939).

However, it is a relief to discover that not all the
configurations are unstable, and Figure 2 shows a group of
four with k = ℓ = 1, α = β = 2, and the same four sets of
values of g and h as in Figure 1. For each of these four
situations, the drift of (x,y), once both dx/dt and dy/dt
are positive, is towards rather than away from the point of
balance, implying a stable end to the process, with a constant
level of armaments (or balance of armaments and trade) on each
side.

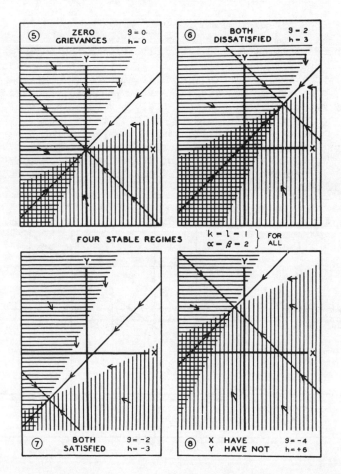

Fig. 2   Four examples of stable configurations with the
         armaments expenditures of the two nations moving
         towards a stable point of balance (from Richardson,
         1939).

The criterion for stability is $\ell k < \alpha\beta$; that is, the
product of the defence coefficients must be less than the
product of the fatigue and expense coefficients. In other
words, the nations must, on average, be more influenced by
the expense of their own defences than by the threat posed
by those of the other nation. Note that the grievances do
not enter into the stability criterion.

APPLICATIONS AND EXTENSIONS

Richardson investigated the application of these general ideas in the first instance to the situation in Europe during the period 1909-14. There were alliances between France and Russia, and between Germany and Austria-Hungary, two groups roughly equal in size. Neither Italy nor Britain was allied with either party. Regarding U and V as the defence expenditures of the two alliances, expressed in the same currency (millions of pounds sterling) and taking k = ℓ, and α = β, Figure 3 shows that a plot of Δ(U + V)/Δt against (U + V) is extremely close to the straight line predicted by Richardson's theory.

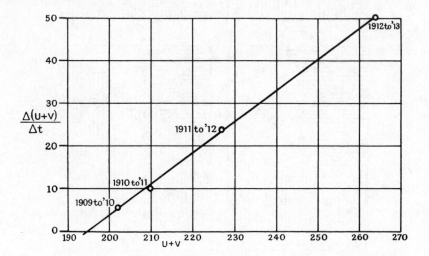

Fig. 3   Annual change in the combined defence budgets of the opposed alliances, plotted against the combined defence budgets (in millions of pounds sterling), 1909-1913 (from Richardson, 1939).

The theory extends readily to three nations, or groups of nations, and so can be used, for example, to explore the relations between two pugnacious nations and a pacifist nation. If the pacifist nation has no grievances against either pugnacious nation, their own aggressive preparations towards it settle to constant levels, while they interact with each other, drifting in some circumstances towards an alliance, in others towards war.

Richardson further extended the theory to n nations, analysed the mathematical properties of the set of n linear

equations, and considered a number of special cases. In
particular he was able to study n nations of unequal sizes
(that is, unequal in their warlike potential), with perfect
communications (that is, with the nations interacting
equally, whatever distance separates them) and with equal
fatigue-and-expense coefficients. This system is stable if
the latent roots of the matrix (which are real) are all
negative. He estimated the coefficients of the matrix for
the nine main Great Powers in 1935, and confirmed that this
matrix represented an unstable situation, the largest latent
root being positive. He then showed that the trend of the
defence expenditures of these nations during the years from
1929 to 1937 was in close conformity with this latent root.

When considering two nations, Richardson also explored the
concept of 'submissiveness', that is, the postulated tendency
of one nation to reduce its defence expenditure when
confronted by a disparity between the defences of another
nation and its own. The concept may be introduced into the
basic equations in several ways. Richardson chose the
following pair of equations:

$$dx/dt = ky \left[1 - \sigma(y - x) \right] - \alpha x + g$$
$$(2)$$
$$dy/dt = \ell x \left[1 - \rho(x - y) \right] - \beta y + h$$

These involve two positive 'submissiveness coefficients' σ
and ρ, applied respectively to the deficit in defences of
each nation, compared with the other. The equations are no
longer linear; some of the resultant possibilities are shown
in Figure 4, assuming zero grievances. With equal
submissiveness coefficients, there may either be stability at
the origin (Diagram 13), or instability there, with
(apparently) two 'whirlpool' points of stability (Diagram 11).
Diagrams 12 and 14 illustrate two possibilities with unilateral
submissiveness only, with either an unstable or a stable point
of balance.

All the foregoing exposition and application of Richardson's
theories is to be found in the 1939 monograph, but there were
two important later developments. One of these was
Richardson's post-war re-examination, in Arms and Insecurity,
of the arms race of 1929-39, using revised data for ten
Great Powers. The revised matrix of defence and fatigue-and-
expense coefficients in 1935 still implies an unstable
situation. It also, through another latent root, implies
the existence of two opposed groups of nations in 1935, with
France, Russia, Czechoslovakia and the United Kingdom on one
side, and Japan, Italy and Germany on the other, the USA,

Poland and China not being clearly on either side. The
inference, from data for 1935, of an eventual alliance between
Britain and Russia against Germany is of particular interest
in view of the surprise non-aggression pact between Russia
and Germany in August, 1939, and the equally unexpected
attack of Germany on Russia in June 1941.

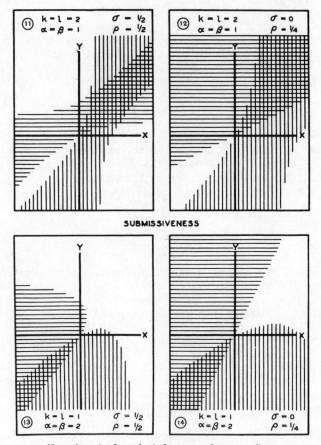

Note: the unit of x and y is four spaces between rulings.

Fig. 4   Two examples of stable, and two of unstable,
         configurations in which one or both nations show
         submissiveness towards the other (from Richardson,
         1939).

The other development concerned submissiveness. Richardson
returned in his last publication (1953b) to this question,
and in particular to the concept of equal submissiveness

between two nations which are nearly equal (in terms of their
defence and fatigue-and-expense coefficients). He investigated
in detail the theoretical consequences of the pair of equations
(2) and of an alternative pair. Both hypotheses were capable
of explaining what usually happens between very unequal
nations - the smaller nation submits. However, for nearly
equal nations the theoretical consequences of assuming that
submissiveness occurred were at variance with observed
historical events. Richardson concluded that he could detect
no evidence of submissive effects between nearly equal nations.

My own experience of mathematical modelling has been in
attempts to derive models for the development of tuberculosis
in an individual following infection, based on the known
immunology and pathology of the disease, and to assess the
ability of those models, for tuberculosis infection and
disease in the individual, to account for various recent
changes in the epidemiological pattern of the disease in the
Netherlands. Some years ago, part of this work was
pre-circulated to a small international group of medical
scientists for discussion at a working meeting. At an early
stage of the discussion the chairman turned to Dr. Kurt Toman,
a respected chest physician from Czechoslovakia, and asked
what he thought of the report. "I think that nothing is
easier to predict than the past". When the laughter had
subsided he explained that he had come across the remark
"nothing is more difficult to predict than the future", and
had turned it upside down. In the process he had transformed
a trite wisecrack into a profound warning to mathematical
modellers. In formulating a theory, and testing it against
observed facts, and especially when modifying a theory which
does not fit observations well, the theory must not be
formulated or reformulated on the basis of the observations.
It may be tempting to tailor a theory in a direction which will
accord with the facts, but there can then be no assurance that
the theory has any real validity. Richardson was very well
aware of this danger, and consistently avoids falling into
any such trap. But he would have enjoyed that remark.

AREAS FOR FURTHER RESEARCH

This brief outline of Richardson's work on international
relations does not adequately convey its quality, in terms
of its scientific and mathematical integrity, nor the
seriousness of its purpose, which was to illuminate our
understanding of the curious ways in which nations react
towards each other, in the hope that thereby it might be
possible to work towards greater international stability.
His theories are not facile, nor ill thought-out, nor are
they simply descriptive of his data. On the contrary, the

accordance of observation with his theories suggests that
those theories may have genuine predictive potential.

There are four main ways in which I think this work can,
and should, be taken further. Some of these have already
been explored, notably the second by the Richardson Institute
for Conflict and Peace Research at Lancaster University.

1. In 1939 Richardson commended to the attention of
    mathematicians the need to investigate the properties
    of the matrix of defence and fatigue coefficients in
    his n-nation theory, when the nations were of unequal
    sizes, but in a world of imperfect communications (that
    is, with the interactions between nations dependent upon
    their geographical separation). His own theoretical
    treatment had been limited to perfect communications.

2. The application of the multi-nation theory to the
    fluctuating events of the past forty years should be
    investigated, with particular reference to the stability
    or instability inherent in the situation at various
    times, and to its agreement (or not) with trends in
    measures of 'threats minus co-operation'. The success
    of this approach, as an analytic technique applied by
    Richardson to the arms race leading up to the second
    world war, is impressive. The collection of reliable data
    for such a long period, even for the major nations, is a
    matter of considerable difficulty, but the present
    availability of computers greatly facilitates analysis
    of the data, once collected. Richardson himself was able
    to cover only three years of the third arms race, namely
    1949 to 1951 (Richardson, 1953a).

3. It is important to consider whether international
    relations are now different in ways which require the
    investigation of other models, or other variants of the
    original model. Three examples follow:

    a) It is often said that the international situation was
        completely changed by the fission bomb, and completely
        changed again by the fusion bomb. I am not convinced
        myself that these were qualitative changes, and not
        simply changes in magnitude. However, the concept of
        the nuclear deterrent, which is possessed by some
        nations and not others, is often advanced as the
        agent responsible for 'keeping the peace (on a world
        scale) for forty years. In Richardson's studies
        submissiveness between nearly equal nations did not
        appear to be an influence which affected their
        behaviour. What about investigating the implied

consequences of the similar but not identical deterrent
proposition?

b)   The effects of using nuclear weapons will not be
     confined to the territory against which they are
     directed, but may, like gas in the First World War,
     also fall upon those who use them, and additionally
     on non-belligerents. The extreme example of this is
     the prospect of a nuclear winter. How may the
     reciprocal implications of the use of nuclear
     weapons, and how may the universal implications,
     affect the behaviour of nations? And is the prospect
     of a nuclear winter different in kind or only in
     degree from the prospect of reprisals? These points
     too require theoretical examination.

c)   Then there is the question of saturation. The
     stockpiles of nuclear weapons are now far greater
     than is required for military supremacy in a nuclear
     war. How may this affect the relationships between
     nations? Does this represent a situation in which the
     fatigue and expense coefficients could become
     collectively greater than the defence coefficients and
     so encourage the development of a stable configuration
     instead of a continued arms race?

4. Richardson hinted in 1939 at the possibility of deliberate
   intervention, which would enable the nations to move
   towards a stable situation from something more finely
   balanced. Essentially it was a proposal to estimate the
   defence coefficients, given the fatigue-and-expense
   coefficients, so as to make the largest latent root of the
   matrix equal to zero; this would represent a just-stable
   situation. It would however only be worth doing this
   against the background of the more realistic theoretical
   model, for unequal nations in a world of imperfect
   communications, referred to above. This might permit a
   theoretical assessment, for example, of current proposals
   for a nuclear freeze. Is such an intervention (assuming it
   could be achieved) likely to represent a move towards
   stability, or the reverse?

During the thirties there was a catch-phase used to
describe the paradoxical behaviour of governments - 'if you
want peace, prepare for war'. Quite recently there was a
cartoon in Punch, showing a couple of white men, with rifles,
encountering a couple of black men in the jungle. The caption
read: 'If you are peace-lovers, why aren't you armed?'. This
suggests that our thinking, and our international attitudes,
have not changed much during the last half-century; if

Richardson had a message then, he has the same message today. He emphasised in 1939 that his equations were

"merely a description of what people would do if they did not stop to think. Why are so many nations reluctantly but steadily increasing their armaments as if they were mechanically compelled to do so? Because, I say, they follow their traditions which are fixtures and their instincts which are mechanical; and because they have not yet made a sufficiently strenuous intellectual and moral effort to control the situation. The process described by the ensuing equations is not to be thought of as inevitable. It is what would occur if instinct and tradition were allowed to act uncontrolled."

The development of the theory so far does not provide a clear indication of the policies or actions, unilateral or joint, that nations should take, and that are likely to be successful, in the attainment of stability in their armaments and in their relationships with each other. Richardson has made a start; and it is my hope that somewhere there is another mathematician or physicist, perhaps even another meteorologist assessing the probabilities of a nuclear winter, who will turn his attention to the wider implications of his subject, and build upon the foundations laid by Lewis Richardson.

REFERENCES

1. Richardson, L.F., 1935, Nature, **135**, pp 830.

2. Richardson, L.F., 1939, Generalised foreign politics. A study in group psychology, British Journal of Psychology Monograph Supplement, **23**, pp 1-91. Cambridge University Press, London.

3. Richardson, L.F., 1953a, Sankhya, **12**, pp 205-228.

4. Richardson, L.F., 1953b, British Journal of Statistical Psychology, **6**, pp 77-90.

5. Richardson, L.F., 1960a, Arms and Insecurity. A mathematical study of the causes and origins of war. (Ed. Rashevsky N. and Trucco E.), pp v-xxv, 1-307. Stevens, London.

6. Richardson, L.F., 1960b, Statistics of deadly quarrels. (Ed. Wright Q. and Lienau C.C.), pp v-xlvi, 1-373. Stevens, London.

SIMULATION OF INTERNATIONAL CONFLICT

P. Smoker
*(Richardson Institute for Conflict and Peace Research,
University of Lancaster)*

1. INTRODUCTION

During the last 25 years conflict research has emerged not
as a discipline in the traditional academic sense, but rather
as a set of approaches to the analysis of conflict. There is
not as yet a broad consensus as to its content or methodology
and indeed, it can be argued, any such agreement would be
premature. This paper traces the development of one perspective
on conflict research and considers the extent to which it
constitutes a viable perspective.

The particular view adopted in this paper is associated with
a "social scientific" approach to conflict. It can be argued,
with some justification, that such "scientific' studies for the
most part have been singularly ineffective in illuminating
our understanding of the complex multilevel processes involved
in many conflict situations. While scientific method has
apparently contributed much to human knowledge in other areas,
it can be argued that human conflict is more amenable to those
types of intellectual skills traditionally associated with the
student of politics.

There are, of course, other viewpoints. Many who have
adopted a "scientific" perspective have done so precisely
because of alleged limitations of purely verbal analyses.
Mathematical and empirical methods can, it is argued, reveal
patterns and processes that may very well elude verbal analyses.
Clearly the large number of scientific studies of conflict,
particularly in the United States, suggests there are many who
do not share the well-rehearsed criticisms of such approaches.

It is not, however, the purpose of this paper to argue the
relative merits of scientific as against traditional modes of

conflict analysis. While it is certainly possible to criticise
both perspectives, this paper focuses on a set of inter-related
studies that, it can be argued, constitute a basis for conflict
research. This is not to argue that this particular
perspective has more or less validity than others, but to
suggest that recent advances in computer technology may provide
fresh impetus for work involving complex conflict systems.

The paper begins by considering game theory and "rational
actor" models. Many traditional theorists have expressed
considerable doubts as to the viability of game theoretical
constructions in the study of complex phenomena. It is argued
by some that models based upon the concept of unitary actors
deciding preferences are too simple to consider complex socio-
political organizations and the complex decision making involved
in such large organisations. This paper does not deal at
length with such criticisms but discusses such "rational actor"
models within the context of a broad range of studies in
conflict theory. In contrast to such decision making analyses,
the second part of the paper considers structural conflict
theorists who locate the problems and solutions to destructive
conflict not in decision making per se but in the very social
and political structures that provide the theatre for decision
makers. Their emphasis on consequences of various structural
formations is in sharp contrast to those concerned with
analysis of particular decisions. The paper discusses some of
the contradictions and dilemmas associated with these differing
perspectives.
Having thus set the stage, later sections of the paper trace
the evolution of mathematical and statistical analyses of
conflict from earlier relatively simple model building to later
complex models.

The final section of this paper deals with prospects for the
simulation of international processes against the background of
the new technology. It is possible that the development of
computer literacy among social scientists is likely to enhance
the intellectual capabilities required for the use of
simulation in the study of complex socio-political systems.
A new generation of researchers accustomed to interactive
heuristic model building may very well be better equipped,
both technologically and intellectually, for constructing truly
systemic models of global phenomena.

But before considering some of the implications of
contemporary and conceivable computer technology it is
necessary to trace the progressive development of "scientific"
approaches to the study of conflict since, this paper argues,
the symbiosis between complex conflict theory and the new

computer technology may very well provide a creative confluence for the simulation of international processes.

2.   RATIONAL ACTOR MODELS

Rational actor models understandably have received a great deal of attention in the literature, particularly since Rapoport published Fights, Games and Debates [55]. His work on experimental games [56] coupled with his contributions to conflict theory [57] make Rapoport a leading exponent of this school of thought. Much of his work has been concerned with the prisoner's dilemma [59] [64], although his review articles explaining the central ideas of game theory have also become standard works [58] [65]. This is not to say that Rapoport is unaware of the limitations of this approach [60], rather to emphasise the extent to which his work has been central to its evolution. Rapoport has been particularly concerned with the relevance of such models to the problems of disarmament [61] while at the same time developing its broader applications to conflict theory [62] [63]. Many others have used game theory to study conflict at various levels, from individual to group to national, across a broad range of conflict situations.

The Prisoners' Dilemma game has been a major focus for this school of conflict analysis. Representing, as it does, a mixed motive situation in which both conflict and cooperation are possible, it has proved of considerable interest to experimental psychologists. The Journal of Conflict Resolution has published scores of studies and a substantial body of experimental findings is now available.

Criticisms of this school have for the most part concentrated on the problems of measuring utility, and on the substantial over-simplifications involved in such analyses. It is not clear how highly simplified laboratory experiments can tell us about the extremely complex real world of international politics, for example, a world in which actors are not necessarily rational, whatever that may mean, and in which the utilities of the various possible outcomes are not clearly defined. Hypergame analysts argue that these criticisms can in part at least be overcome since their whole definition of the "game" is a subjective matter.

Rapoport and others have counter argued that there is a heuristic value in Prisoners' Dilemma type experiments that in and of itself justifies such simplified conflict models. Certainly by illustrating that so called rational decision making, where an actor is primarily motivated by self interest, can produce outcomes clearly at odds with the same self

interest, the actor theories of conflict are at least in
general terms relevant to an important set of problems. Much
of the academic and public discussion of theories of deterrence
is couched in similar conceptual terms. The Soviet Union, The
United States are discussed as if they were individual rational
actors deciding policy, and arguments for or against the
deterrent are often justified in terms of rational choice.

In many ways, the most compelling arguments against rational
actor models stem from the problems of extending ideas that
might, under some circumstances be appropriate for a
hypothetical rational individual to analyses of complex human
situations. Clearly the United States and the Soviet Union
are complex economic, political, social structures and
decision making, be it individual or collective, within and
between such structures is likely to involve numerous non
rational factors, as well as, possibly, some that might be
referred to as rational. The advantage of simple rational
actor models is that they explore, even be it in artificial
isolation, some of the consequences of rational decision
making, and some of the deviations from such decision making
by individuals in laboratory situations. The overwhelming
weight of experimental evidence, as one might have expected,
shows a variety of individual differences with regard to
actual decision making in laboratory situations. The
corresponding theoretical conclusions regarding rationality,
as for example in the prisoners' dilemma, suggest fundamental
inadequacies with simple definitions of rationality and the
need to develop broader frameworks to allow for such
possibilities as the interaction of utilities, which allow for
the interdependence of parties in conflict situations.

An important, though perhaps unintended, consequence of
rational actor theories is to raise substantial doubts as to
the viability of conflict analysis, be it academic, political
or popular, if such analysis in some way or other assumes
predominantly "rational" criteria, however rational may be
defined, to be central. Subsequently this paper will develop
this idea as a part of the particular systems perspective it
hopes to elaborate. But before this is possible a second
component of this perspective must be reviewed, however briefly.

3. STRUCTURAL CONFLICT THEORIES

If Anatol Rapoport has been the champion of the rational
actor school, then Johan Galtung has similarly been the dominant
figure in arguing for structural conflict theories. In many
ways the Journal of Peace Research has performed for the
structural theorists a similar role to that played by the
Journal of Conflict Resolution for those with an "actor"
persuasion.

Galtung of course was able to build on a long tradition of structural theorists. Coser [9] during the 1950's had developed his theories concerning the functions of social conflict, Parsons [52] had elaborated his conservative conception of the Social System, and Merton [49] had contributed considerably to the idea of a social structure [10]. But within contemporary conflict research Galtung has emerged as the leading structural theorist. During the early sixties a string of papers developed structural analyses of aggression [16], foreign policy opinion [17], institutionalised conflict resolution [18], east west interaction patterns [19], the future of the International system [20], integration [21], the structure of foreign news [24], and patterns of diplomacy [25]. Many others elaborated and explored Galtung's varied structural hypotheses.

Conflict researchers during the early sixties were, for the most part, concerned with the arms races, violence, war and aggression and structural theorists under Galtung's intellectual leadership established an important, not to say dominant, school of research, particularly in the Scandinavian countries. It is perhaps no accident that rational actor models with their emphasis on individual decision makers for the most part developed in the United States at that time while the structural school, with its emphasis on the importance of the social structure, blossomed in Europe. The late sixties, however, saw a sudden turning point for conflict research, a turning point that, on the face of it, was associated with the Schmidt-Galtung debate.

Schmidt published Politics and Peace Research in 1968 [75], a paper in which he argued vigorously against the then dominant perspective of structural theorists. His critique began by considering three sets of problems. To begin with he argued that in most cases the values implicit in peace research support the maintenance of the existing order. Secondly Schmidt questioned the then dominant neutral, objective or impartial stand attempted by many researchers who similarly adopted a symmetrical position in considering the structure of a conflict, regardless of whether or not this was actually the case. And thirdly Schmidt argued that the institutionalisation of peace research coupled with the notion that the findings of peace researchers could be used by decision makers to help control the international system, meant that peace researchers would have to ally themselves with those who have power in the international system to get their policy proposals implemented. Further, Schmidt argued, since the powerful want to maintain the stability of the international system, from which they benefit, peace research had at that time adopted a similar value position. Such a

value position would maintain inequalities, injustice and
exploitation provided a relatively stable war free state
existed. Schmidt argued that the most important point running
through all these problems was that negative peace, that is
the absence of war, was the only value that had sufficient
consensus among the dominant style of peace research. Since
there was no consensus on positive peace the problems
according to Schmidt were almost inevitable. From the
structuralists' point of view Schmidt, in one significant way,
took a more extreme position than many of his contemporaries.
Whereas most structuralists accepted what Schmidt called a
"subjective" model of conflict, in which the conflict, defined
in terms of incompatible values or goals, the conflict
behaviour and the conflict attitudes all interact, Schmidt
argued for an "objectivist" position where conflict is not
seen as a matter of subjective definition but is embedded in
the social structure itself. This "objectivist" model clearly
implies that conflict can only be resolved by changes in
structure. The conflict may be latent, for example a
satisfied slave, but such conflicts, Schmidt argued, should be
dealt with by changing the structure. Up to that time the
majority view amongst structuralists in conflict research was
that polarization for example was a conflict situation to be
avoided and overcome through procedures such as introducing
positive links, changing attitudes or modifying behaviour.
For the "objectivist" view of conflict the conflict is
embedded in the social structure and can only be changed by
changing the structure through polarisation, making the latent
conflict manifest, and struggle.

    It is of some interest that Schmidt singles out the
Prisoners' Dilemma paradigm to illustrate his point. His
critique of this school is not concerned with measurement of
utility or problems of complexity, rather that experiments
such as those undertaken with the Prisoners' Dilemma where
participants have to work within the given structure of the
game reflect the dominant view of society in which change of
structure is rejected or not allowed and solutions must be
found within the existing structural constraints. Subsequently
the idea that the Prisoners' Dilemma game can be seen as a
system of social domination and that effort should be directed
at changing the structure rather than working with the given
possible outcomes has been developed [48]. Schmidt's advocacy
of revolution research and the use of conflict as a procedure
for structural change led him, understandably, to advocate
abandoning the term peace research, a less than desirable
activity within his framework.

Galtung's response [22], it can be argued, changed the course of
structural conflict theory for the ensuing decade. By
broadening the definition of violence to include both physical
and structural violence, and by defining negative peace as the
absence of physical violence and positive peace, or social
justice, as the absence of structural violence, he developed
the argument that peace research can and should consider both
types of violence and the relationship between them [26]. The
focus for structural conflict theories had also changed and
studies of dominance systems [44], dependency structures [27]
and the imperialism [51] replaced studies of war, violence and
aggression as central topics for structural conflict theories.

It does not seem unreasonable to argue that the Schmidt-
Galtung debate was the critical factor in changing the
direction of structural conflict theory. But it is of interest
to observe that comparable changes were taking place in the
broader society. During the late fifties and early sixties
the anti nuclear weapons movement began, peaked, and subsided,
and the anti Vietnam war movement became the main focus. With
the ending of United States military involvement the broader
peace movement became concerned with ecology, nuclear power,
women's liberation and a range of issues not concerned with
war or physical violence so much as with questions of
structural violence and quality of life. Under such broader
societal conditions the shift of emphasis within conflict
research is not altogether surprising.

It can be argued that during the late sixties and the
seventies an over-reaction took place among some academic
conflict researchers that resulted not only in an undue
emphasis on structural violence as against physical violence
but also in the neglect of a line of argument that contained
some promise, particularly for those concerned with the
simulation of international processes. The next section of
this paper deals with that argument.

4. THE ENTROPY DILEMMA

Schmidt, prior to publication of his article in the Journal
of Peace Research, had presented his ideas at Galtung's then
home base the International Peace Research Institute in Oslo
during January 1968. At that time Galtungs own paper, Entropy
and the General Theory of Peace [23], which he had presented
four months previously at the second International Peace
Research Association conference, was still at the printers.
It is of some interest that this paper, a theoretical
construction that probably excelled all of his previous work
in its brilliance, complexity and scope, is amongst his less
well known works. Such was the fascination of the structural

violence concept that the theoretical framework put forward by
Galtung in the Entropy paper was not developed and, in a sense,
marked the end of a line of conflict theory that certainly
contained possibilities for development.

Galtung in this paper argued that the concept Entropy could
be applied to social systems, an assumption that many no
doubt would disagree with. For Galtung a minimum entropy
position for any property would be represented by a point
distribution, for example all the guilt in a particular
situation being ascribed to one actor, while a maximum entropy
position would assign guilt across all parties using say a
uniform distribution. Galtung in fact anticipated Schmidt's
subsequent criticism of the symmetry of the peace research
perspective by recognising that the maximum entropy position
in this case "lays peace research open to the accusation that
it has a built-in symmetry in its way of dealing with any
pairs of parties to a conflict, regardless of whether the
conflict is between David and Goliath or between two Davids or
two Goliaths."

Galtung argues that actor entropy and interaction entropy
are fundamental to social science. Actor entropy is based on
the distribution of actors on positions, for example power,
wealth or other rank dimensions, and is clearly related to
much structural analysis. Interaction entropy is based on
the distribution of links between actors, for example trade
between nations or information flows between groups.
Interaction entropy is a systemic concept that focuses on
links between various actors and variables, a critical
problem for those involved in the simulation of international
processes. A social system high on both actor and interaction
entropy, according to Galtung, would have a wide range of
different actors with a wide variety of links distributed
fairly evenly among them. A social system low on both actor
and interaction entropy would have actors heavily distributed
in one or a few positions and interaction limited to just a
few of the many possible links. Understanding the pattern of
interaction entropy is a crucial problem for complex conflict
models if wholistic system simulations are to more adequately
represent the complex dynamics of large socio-political
systems.

Galtung uses actor and interaction entropy to examine a
variety of alternative world models and to elaborate the
properties associated with various levels of actor and
interaction entropy. Galtung argues that in low entropy
systems, where actors and interaction are distributed across
relatively few positions, conflict will be less frequent than
in high entropy systems but more damaging and intense when it

does in fact break out.   "In a high entropy system" he argues, "the world is more complicated, more "messy" as we have expressed it.   The possibilities at any point in the system are more numerous; the range of interaction patterns and chains of interaction much broader.   Conflicts are absorbed locally; they may be numerous indeed but their consequences are slight."

Galtung postulates that the relationship between conflict and entropy is fairly clear.   In the high entropy state, with actors distributed across a wide range of positions and interaction broadly distributed across the maximum number of links, there will be   numerous micro conflicts at the local level but macro conflict is very unlikely since the messy structure does not permit it.   In the low entropy state micro conflicts are much less frequent but macro conflict becomes more likely since the energy in the system can become concentrated at particular points.   Perhaps, without realising it, Galtung in his entropy paper had predicted and in fact answered Schmidt.   The complexity of the entropy paper, and the natural antipathy many individuals have to the use of the idea of entropy in the human context, may have led Galtung to answer Schmidt in the way he did [22].   Indeed it is doubtful that those who would have been swayed by Schmidt's case with its manifestly political argument would have understood fully the structural dilemma of the entropy model, a dilemma that is at least as intellectually interesting as that of the prisoner in game theory.

The low entropy model, according to Galtung, has a consequence the possibility of major, and perhaps disastrous conflict.   The high entropy alternatives avoid this problem but make macro changes much less likely, including changes in global inequalities and situations of "structural violence".   The energy in high entropy states manifests itself in local conflicts and micro changes.   Galtung points out that "Our fear of the consequences of the bad purposes would make the peace theorist focus his attention on the high entropy systems – but he should be aware of the fact that such systems will have less macro energy at their disposal, that the energy will be stored in small, local units so to speak, so that changes will tend more to be micro changes.   And this means that those who feel that there are still some macro changes to be carried out will feel the time has not yet come for major entropy increases, that first there must be a basic dissociative split that can be used for a major transformation, and then, afterwards, the time will be ripe."

The entropy dilemma for the structural and systems theorist presents an intriguing challenge, a challenge that has not received attention comparable to the energies applied to the

Prisoners' Dilemma. The problem involved in creating global
systems that are not vulnerable to macro conflict in general,
and nuclear war in particular are difficult enough. When the
same systems have in addition to be capable of sufficient
structural transformation to solve the numerous macro problems
associated with global inequalities the difficulties are
compounded. Those involved in the simulation of international
processes clearly face considerable difficulties in developing
such models, but given the complexities of the interlocking
systems involved, the arguments for such an approach are
compelling. The last section of this paper considers these
arguments in more detail.

5. MATHEMATICAL MODELS

Rational actor studies, particularly those associated with
the theory of games, owe much to mathematical models [86], but
has been a wide variety of mathematical models used in the
study of conflict. Richardson [68] was one of the first to
develop this style of conflict research with his simple
mathematical models of arms races. There are, of course, many
problems with such studies that can be illustrated by
considering Richardson's work. Firstly the assumptions upon
which any mathematical model is built are open to question. In
Richardson's case, for example, an arms race is assumed to be
caused primarily by interaction effects between two or more
parties, although it is not always realised that his original
equations contain an interaction term through the defence
coefficient, and an autistic (internally-generated) term
through the fatigue and expense coefficient. This is
discussed in some detail shortly. There are various other
hypotheses, such as the consequences of a military industrial
complex or pure autistic or interaction theories that would
lead to a completely different model. Secondly the
oversimplification involved in developing such models, as for
example in the case of game theory, can, it is argued, lead to
inadequacies. The criticism of Prisoners' Dilemma on this
point is now well rehearsed. And thirdly the problem of
measuring variables used in a mathematical model can severely
limit the applicability of the model. In Richardson's case,
for example, "defences" are not easily quantifiable and a wide
variety of possibilities exists for indexing such a concept.
It is not always understood that verbal theories too face
similar problems. The assumptions can be questioned, they are
almost inevitably oversimplified, and the concepts may or may
not correspond to real world phenomena.

An important contribution that mathematical models have made
to the study of conflict and to the use of simulation concerns
the development of an innovative and experimental attitude.

The last section of this paper argues that this consequence of
the growing computer literacy among social scientists will be
an important contribution to global systems models. <u>Provided
models are interpreted as models and not as realities, they can
help to illuminate our understanding of particular phenomena</u>.
In many ways the use of models requires an acceptance that
ultimately all models are probably false, but that some are
less false in particular ways than others. The primitive state
of development of social science makes it much easier to accept
the inadequacies of contemporary mathematical models and to use
them in a heuristic fashion. It is much harder to accept that
Einstein's model is only a model and that one day it is likely
to be falsified or replaced by a less false representation.

When Richardson first began his research some seventy
years ago, the idea of applying mathematical techniques to the
study of arms races or international processes was novel.
Today these approaches, along with a variety of others, have
helped to develop a perspective on global systems that is
likely to evolve further in the future. Richardson's models,
understandably, are today seen as over-simple and not
necessarily applicable to the contemporary situation, involving
as it does a wide variety of possible complicating factors.
Nevertheless the emerging systemic perspective on international
processes owes much to Richardson's early efforts. His
equations, because of their interactive component, operate as
a system in which interdependence is critical and in which all
actors are, to a greater or lesser extent, influenced by the
total context.

Richardson's mathematical model of an arms race has become
a classic paradigm for escalation processes. It is often
argued that Richardson's equations illustrate interaction
aspects of escalation as distinct from "autistic" models that
stress domestic pressures driving arms races, pressures that
are quite distinct from rivalry or balance of power
considerations. This paper takes a different view and suggests
that the classic Richardson models contain both interactive and
autistic components as well as the mathematical term associated
with exogeneous variables. These ideas are developed to
present a systems framework that could be used to consider
various aspects of processes capable of leading to accidental
nuclear war.

The classical Richardson model argues that if x represents
the "defences" of nation A, and y the "defences" of nation B
then

$$dx/dt = k\,y - a\,x + g \qquad (1)$$

$$dy/dt = l\,x - b\,y + h \qquad (2)$$

where k and l are defence coeffients, a and b are fatigue and
expense coeffients and g and h are constants. The first
terms in each of these equations, k y and l x, are interaction
terms influencing the rate of change of each sides defences
(dx/dt and dy/dt) as a result of the level of others "defences".
The defence coefficients k and l can be interpreted as
sensitivity to the other sides so-called defences ("defences").
The higher k and l, the greater the sensitivity, the greater
the interaction effect between the two countries. If k and l
are set at zero, then the equations describe a non-interaction
(autistic) situation. In fact Richardson used such a procedure
to estimate the size of the autistic coefficients a and b by
considering the period immediately after the second world war
when, he assumed, k and l were zero and the annual decrease
in defence expenditure was associated with the various internal
problems of changing from a war time to a peace time economy.

These autistic terms a x and b y are used by Richardson to
represent "fatigue and expense", the assumption being the
greater the expenditures the more one's own defences are
limited as a result of cost. These terms represent economic
constraints which are entirely internal to the nation
(autistic).

An alternative interpretation of the Richardson model would
class all autistic factors together as follows.

$$dx/dt = k y + a x + g \qquad\qquad (3)$$

$$dy/dt = l x + b y + h \qquad\qquad (4)$$

The constants a and b then represent the net consequences
of internal autistic factors, economic and otherwise. Here the
convention is that if the autism coefficents a or b are
positive then the net consequence of such factors is
escalatory, and if they are negative, as Richardson assumed,
a decrease results. If a and b are zero we have a pure
interaction model with no autistic component, and if the
interaction coefficients k and l are zero we have an autistic
model with no interactive component. The relative size of k
and l as compared to a and b is an indicator of the relative
importance the model gives to interaction as against autism,
while the sign of the coefficents (positive or negative)
indicates escalatory or dampening consequences of interaction
and autism.

The final terms g and h were seen by Richardson as grievance
terms related to treaties and trade. For the purposes of this
paper the terms g and h represent exogenous factors independent
of interaction or autistism. These exogenous factors, as we

shall see, may be a consequence of interaction or autistic behaviour outside the particular sub system described by the equations (3) and (4).

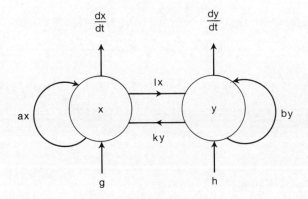

Fig. 1

Figure 1 illustrates the basic features of the simple three factor model developed from Richardson's equations. Here the behaviour dx/dt is a result of the interaction k*y, the autistic effect a*x, and the exogenous input g, while dy/dt results from the interaction effect l*x, the autistic effect b*y and the exogenous term h.

It is not necessary to assume that the three sets of coefficients are constant. Richardson was forced to make this assumption because of the computational equipment available to him and his need to generate pure mathematical solutions to the equations. The systemic development of these assumptions accepts that such short term "constants" can change as a result of changes in the overall situation. For example it can be argued that in an escalating arms race or crisis situation sensitivities to armaments increase.

An important problem for mathematical models of human behaviour is whether relationships are seen as constant, a perspective clearly adopted by Richardson, or subject to change. This paper adopts the latter view and argues that relationships are represented by "constants" such as k,l and a,b are themselves subject to change. Discussion of a complex systems model will illustrate the point.

Richardson's simple model has been used by a number of authors as a basis for considering macro aspects of arms races. But in order to consider complex conflict situations a comparatively complex model is almost certainly more

appropriate. Such a model has been used to study complex
conflict situations and is based on interlinking simple
subsystems of the type illustrated by equations (3) and (4) in
figure 1.

To begin with let us consider two such subsystems

$$dx/dt = k\ y + a\ x + g \quad (3) \quad dy/dt = 1\ x + b\ y + h \quad (4)$$

$$du/dt = m\ v + c\ u + i \quad (5) \quad dv/dt = n\ u + d\ v + j \quad (6)$$

These subsystems may be completely independent as illustrated
in figure 2.

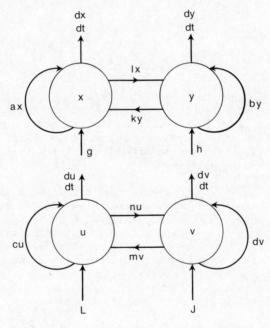

Fig. 2

Or they could be connected in a variety of ways. Supposing
for example that

$$g=du/dt \quad (7) \quad h=dv/dt \quad (8) \quad i=dx/dt \quad (9) \text{ and } j = dy/dt \quad (10)$$

we would then have the situation illustrated in Figure 3.

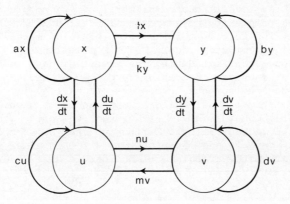

Fig. 3

From a whole systems perspective Figure 2 illustrates the case of two autistic sub systems with unrelated exogenous variables, that is a whole system comprising two subsystems that are completely independent of one another, while Figure 3 illustrates an interactive whole system with no exogenous components. The system in Figure 3 includes autistic components a*x,b*y,c*w and d*v. If these terms were removed the system would be entirely interactive. These two extreme cases, an entirely interactive whole system and a whole system comprising independent subsystems, bracket numerous in-between situations in which subsystems to some extent interact, to some extent operate as autonomous systems, and to some extent are influenced by factors outside the whole system.

Almost inevitably, more complex mathematical constructions have developed in the study of conflict, particularly in computer simulation. During the fifties and early sixties Guetzkow [30][31] was developing the Inter Nation Simulation, and Benson his simple Diplomatic Game [3]. These early primitive models were used by researchers for a variety of purposes including attempts to simulate the outbreak of World War One [37] ,[54] , an examination of the possible consequences of nuclear proliferation [6], and investigations of deterrence theories [66] [11]. Researchers such as Hermann applied simulation to the study of crises [38][39], while others explored the value of simulation as a teaching device for international relations [32][70].

During the sixties the mathematical component of simulations of international relations increased [31] and more complex constructions emerged [33] [45], including all-computer models

such as TEMPER [1] and the work of the Club of Rome [15] [50].
While earlier models focused primarily on macro processes, the
increasing complexity of computer simulation approaches
inevitably led to computer simulated decision makers immersed
within larger computer models [5]. Simulations of
international conflict during the seventies have seen a
substantial increase in complexity [14] [34] and with the
significant developments in computer technology it is likely
that mathematical model building approaches to global systems
will benefit considerably. Richardson, equipped with log
tables or a mechanical calculating machine, was not able to
experiment quickly and interactively with his model whereas the
contemporary mathematical model builder is able to explore the
properties of increasingly complicated models in a creative and
immediate way.

The systemic perspective is likely to benefit considerably
from such developments. The idea of conflict systems may yet
replace the idea of actors or conflict structures as a central
focus for research.

6.   STATISTICAL ANALYSES

A final input into the emerging systems perspective on
conflict analysis comes from statistical approaches. Authors
such as Wright [90] and Sorokin [84] in their definitive works
present a range of statistical evidence concerning patterns in
warfare and violence. Richardson [69] also developed a
statistical approach to the study of war that has subsequently
been developed further by Singer [82] and others [12] [13]   In
fact multivariate statistical analyses came relatively early in
the gradual evolution of the global systems
perspective [71] [85] . At first such analyses were cross
sectional, that is they did not consider the dynamics of
systems but were rather snap -shots of the structure at a
particular time.

During the sixties some simple time series analysis of
conflict did take place by researchers such as Alker [2],
McClelland [47] [48] [81] and Holsti [42] and the limitations
of cross sectional analysis, like the limitations of static
structural analysis, became more obvious. With the development
of the idea of systems and the increasing power of computer
technology attempts at multivariate time series analysis of
conflict phenomena became a feature of the late seventies with
studies such as those of Wright [89] being typical. Such
studies see conflict systems as dynamic entities that do not
necessarily have fixed patterns. Whereas it is possible to
conceive of systems as having relatively fixed dynamics, it is
also possible to envisage complex conflict systems, such as

that studied by Wright in Northern Ireland, as systems
containing actors and structures with both autistic and
interactive processes, the processes themselves being subject
to change as a result of the overall conflict dynamics.

An Example

A multivariate time series analysis package has been
developed at the Richardson Institute based upon the assumptions
outlined above. The purpose of the package is to develop a
picture of a whole system and its changing dynamics over time.
Figures 4, 5 and 6 illustrate some of the results of an
analysis of the relationships between a variety of indicators
of the conflict in Northern Ireland.

Figure 4 illustrates the dynamics of the Northern Ireland
conflict before the introduction of internment (a state of
emergency that allows imprisonment without being charged with
an offence for a period of up to 48 hours). The convention
used in the diagrams is similar to that discussed above except
the lines linking variables (circles) can have a directional
arrow. A line without an arrow indicates interaction, that is
two variables rising or falling together at a particular
point in time. A single arrow indicates a lag of one month in
the linkage between the two variables, and a double arrow a lag
of two months. The direction of the arrow indicates earlier
to later.

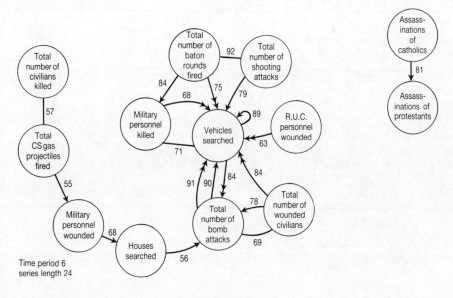

Fig. 4

In Figure 4 we see a fairly weak conflict system, with
relatively few linkages and one pair of conflict indicators,
the mutual assassination of catholics and protestants, forming
an independent conflict system. The variable "vehicles
searched" is almost a conflict sink since output from six other
conflict indicators feeds into it. There is however feedback
from this variable to military personnel killed and to the
total number of bomb attacks. The Governmental forces at that
time were clearly searching vehicles as a response to a wide
range of activities by opposing non-governmental groups (the
IRA and UDA) and these groups in turn were responding to these
searches by killing military personnel and instigating bomb
attacks.

The numbers against the line give an indication of the
level of association, zero being low and 1 being perfect
association. They can be interpreted as correlation
coefficients, although tests of significance are not
appropriate.

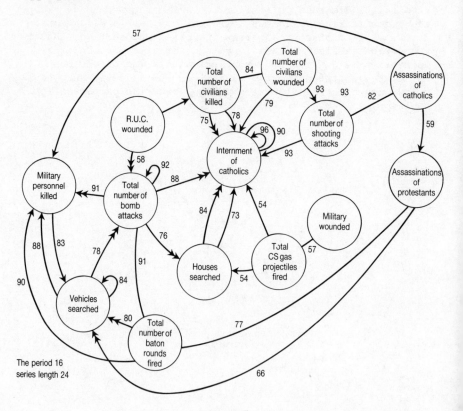

Fig. 5

Figure 5 shows an increase in intensity of interaction as
the conflict escalated with the introduction of internment.
Internment of catholics has clearly become a conflict sink and
all the conflict indicators have become part of one conflict
system. This phenomenon is common in escalating conflicts and
crisis situations, relationships intensify, and multiply and
the system as a whole becomes important. A system of the sort
illustrated in Figure 5 does not necessarily behave in easily
predictable ways. The increased intensity is associated
with changed patterns between variables. Figure 6 illustrates
the different patterns that had established themselves much
later in the conflict when internment was less important
(dotted lines indicate negative relationships).

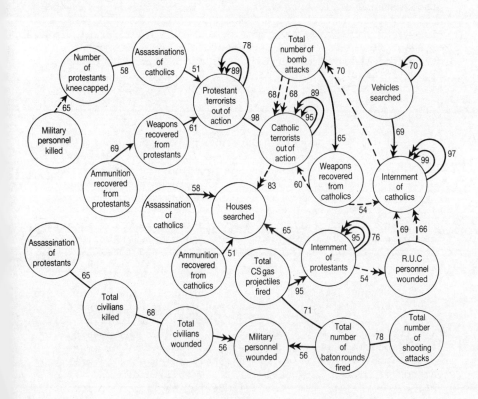

Fig. 6

7. SYSTEMS SIMULATIONS: THE NEW FRONTIER

This paper has developed a pattern of argument that suggests
that the evolving systems perspective on the analysis of global
processes is well-matched to take advantage of the recent
developments in computer technology. While the idea of systems
is certainly not new, it can be argued that the ability for
systems thinking has not been well developed. The idea of
complex multilevel systems in which relationships change in
accordance with system perogatives has not been implemented
in global models because the conceptual apparatus of simulators
has not been tuned to whole systems. As we move from
"science" perspectives to "systems" perspectives, a move that
will be enhanced by increased computer literacy, our thought
processes and style of analysis will increasingly implement
whole systems models. To a large extent the contemporary
micro computer can be seen not so much as a tool for simulation,
although it certainly can be effectively so used, but as a
device for thought processing. The micro computer, with its
highly interactive visual capabilities has replaced the note
pad or typewriter for many social scientists, and the thought
processes associated with such technology are likely to become
increasingly evident in global systems simulations.

The state of theoretical development in conceptualising
global systems is fortunately poised for further advances. It
can be argued that simulation now provides the only approach
to the types of realistic models that are necessary. When
visionary academics such as Harold Guetzkow first pioneered
simulation of international relations such a claim would have
been bold indeed. But it is now hard to dispute the assertion
that models of global systems should be complex cybernetic
constructions, and that simulation is required to explore their
properties.

The entropy dilemma is a systems problem. The conservative
requirement of stability and the radical imperative for change
present apparently contradictory demands. But it is now clear
that the contemporary global system, in which the possibility
of thermonuclear war coexists with the actuality of large
scale global inequalities, is unlikely to be able to meet the
demands of either perspective. The stability demands for
absence of war and the just demands for abolition of poverty,
disease and macro structural inequalities, can, and must, be
redefined within the context of a wholistic systems perspective.
The possibility that the solution to either problem does not
require the neglect of the other is at the heart of this
wholistic approach.

CONCLUSION

This paper has argued that a number of different approaches to conflict analysis have contributed to the gradual evolution of a systems perspective. The ramifications of this perspective have not, as yet, been adequately explored. For example a rational actor in a conflict system would not necessarily adopt the types of criteria associated with rationality by traditional game theorists. The "unintended consequences" of complicated social systems would require a rational actor to take systems dynamics into account since effective systems behaviour by an actor requires an understanding of the total dynamics.

The idea of wholistic analysis presents considerable conceptual and methodological difficulties, and it is by no means certain that these will be overcome, given the very limited amount of intellectual effort that is likely to be applied to these problems. Nevertheless during the eighties, particularly given the continued advances in interactive computer technology, systemic models of global processes are likely to be an important component of an evolving academic outlook.

These models can increasingly interconnect the seemingly unconnected aspects of global systems. In particular the apparently competing demands of physical and structural violence, of peace and justice, can be reinterpreted within a truly systemic perspective.

REFERENCES

1. Abt, C. and Gorden, (1969) Report on Project TEMPER. in Pruitt, D and Snyder, R, eds, Theory and Research on the Causes of War, Englewood Cliffs, NJ.

2. Alker, Hayward. (1967) Causal Inference and Political Analysis in Joseph Bernd, ed, Mathematical Application in Political Science, Charlottesville, University Press of Virginia.

3. Benson, O., (1961) A simple Diplomatic Game in International Politics and Foreign Policy, edited by Rosenau, MacMillan.

4. Brams, S., (1975) Game Theory and Politics. Free Press, 312pp.

5. Bremer, S, (1977) Simulated Worlds: A computer Model of National Decision Making. Princeton University Press.

6. Brody, R., Some Systemic Effects of the Spread of Nuclear
    Weapons Technology: A Study through Simulation of a
    Multinuclear Future . Journal of Conflict Resolution,7,
    pp. 663-753.

7. Burns, B. and Buckley, W., (1974) The Prisoner's Dilemma
    Game as a System of Social Domination . Journal of Peace
    Research, pp. 221-228.

8. Coddington, A., (1967) Game Theory, Bargaining Theory and
    Strategic Reasoning.   Journal of Peace Research,
    Number 1, pp39-45.

9. Coser, L., (1956) The Functions of Social Conflict.
    Routledge, 188 pp.

10. Coser, L., (1975) The Idea of A Social Structure: Papers
     in Honor of Robert Merton. Harcourt Brace, 547 pp.

11. Crow, W.J., A Study of Strategic Doctrines using the Inter
     Nation Simulation . Journal of Conflict Resolution, 7(3),
     pp. 580-589.

12. Denton, F., (1966) Systemic Properties of War 1820-1949.
     The RAND Corporation.

13. Denton, F. and Warren, P., (1967) Some Cyclical Patterns
     in the History of Violence. The RAND Corporation.

14. Deutsch, Karl, Fritsch, Gaguaribe, and Markovitts. (1977)
     Problems of World Modelling: Political and Social
     Implications. Cambridge, Ballinger.

15. Forrester, J.V., (1971) World Dynamics. Cambridge, Mass.,
     Wright Allen.

16. Galtung, Johan., (1964) A structural Theory of Aggression.
     Journal of Peace Research, Number 2, pp. 94-119.

17. Galtung, Johan. (1964) Foreign Policy Opinion as a
     Function of Social Position · Journal of Peace Research,
     Number 3/4, pp. 206-231.

18. Galtung, Johan., (1965) Institutionalised Conflict
     Resolution . Journal of Peace Research, Number 4, pp. 348-
     390.

19. Galtung, Johan, (1966) East West Interaction Patterns
     Journal of Peace Research, Number 2, pp. 146-177.

20. Galtung, Johan , (1967) On the Future of the International System . Journal of Peace Research, Number 4, pp 375-396.

21. Galtung, Johan., (1968) A Structural Theory of Integration. Journal of Peace Research, Number 4, pp. 375-396.

22. Galtung, Johan , (1969) Violence, Peace and Peace Research. Journal of Peace Research, Number 3, pp. 167-76.

23. Galtung, Johan , (1969) Entropy and the General Theory of Peace. In, Proceedings of the Second Conference of the International Peace Research Association.

24. Galtung, Johan and Ruge, Marie Holmboe, (1965) The Structure of Foreign News. Journal of Peace Research, Number 1, pp. 64-92.

25. Galtung, Johan and Ruge, Marie Holmboe, (1965) Patterns of Diplomacy. Journal of Peace Research, Number 2, pp. 101-135.

26. Galtung, Johan and Hoivik, Tord. (1971) Structural and Direct Violence. Journal of Peace Research, Number 1, pp. 73-76.

27. Gantzel, K.J., (1973) Dependency Structures as Dominant Patterns in World Society . Journal of Peace Research, pp. 203-215.

28. Gleditsch, Nils Peter, (1967) Trends in World Airline Patterns. Journal of Peace Research, Number 4, pp. 366-408.

29. Guetzkow, H., (1959) A Use of Simulation in the Study of International Relations. Behavioural Science 4(3), pp. 183-191.

30. Guetzkow, H., (1962) Simulation in Social Science: Readings. Englewood Cliffs, NJ, Prentice Hall.

31. Guetzkow, H., (1969) Some Uses of Mathematics in Simulation of International Relations in J. Bernd, ed, Mathematical Applications in Political Science, Dallas, Southern Methodist Press.

32. Guetzkow, Harold with Alger, Brody, Noel and Snyder. (1963) Simulation in International Relations: Developments for Research and Teaching. Prentice Hall.

33. Guetzkow, Harold, Kotler and Schultz, eds., (1972)
    Simulation in Social and Administrative Science: Overviews
    and Case Examples. Prentice Hall.

34. Guetzkow, Harold and Valadez., (1981) Simulated
    Processes: Theories and Research in Global Modelling.
    Beverley Hills, Cal. Sage.

35. Halle, Nils., (1966) Social Position and Foreign Policy
    Attitudes . Journal of Peace Research, Number 1, pp. 46-75.

36. Harsanyi, J., On the Rationality Postulates Underlying the
    Theory of Co-operative Games.   Journal of Conflict
    Resolution, Volume V, Number 4, pp. 354-365.

37. Hermann, C.F. and Hermann, M.G. (1967) An Attempt to
    Simulate the Outbreak of World War One.   American Political
    Science Review, pp. 400-416.

38. Hermann, C.F., (1972) Threat, time and Surprise: A
    Simulation of International Crisis in C.R. Herman (Editor)
    International Crises: Insights from Behavioural Research,
    Macmillan.

39. Hermann, C.F., Counterattack or Delay: Characteristics
    Influencing Decision Makers Responses to the Simulation of
    an Unidentified Attack.   Journal of Conflict Resolution,
    18 , pp. 75-106.

40. Hernes, Gudmund, (1969) On Rank Disequilibrium and
    Military Coups D'Etat . Journal of Peace Research, Number 1,
    pp. 65-76.

41. Himmelstrand, Ulf. (1969) Tribalism, Nationalism, Rank
    Equilibrium and Social Structure . Journal of Peace
    Research, Number 2, pp. 146-171.

42. Holsti, Oli. (1966) Comparative Data from Content
    Analysis: Perceptions of Hostility and Economic Variables
    in the 1914 Crisis in Richard Merritt and Stein Rokkan, eds.
    Comparing Nations: The Use of Quantitative Data in Cross
    National Research, New Haven, Yale University Press.

43. Hveem, Helga. (1968) Foreign Policy Thinking in the Elite
    and the General Population . Journal of Peace Research,
    Number 2, pp. 146-171.

44. Hveem, Helga., (1973) Global Dominance Systems.   Journal
    of Peace Research, pp. 319-340.

45. Laponce, Jean and Smoker, eds., (1972) Experimentation
and Simulation in Political Science, University of Toronto
Press.

46. Lucas, R. Duncan and Raiffa, Howard. (1964) Games and
Decisions. John Wiley.

47. McClelland, C., (1962) Decisional Opportunity and
Political Controversy: The Quemoy Case. Journal of
Conflict Resolution, pp. 201-215.

48. McClelland, C., (1964) Action Structures and Communication
in Two International Crises: Quemoy and Berlin.
Background, pp. 201-215.

49. Merton, R., (1968) Social Theory and Social Structures.
Free Press, 702 pp.

50. Mesarovic, M. and Pestel, E., (1974) Mankind at the
Turning Point: The Second Report of the Club of Rome,
Elsevier.

51. Oberg, J., (1975) Arms Trade with the Third World as an
Aspect of Imperialism . Journal of Peace Research,
pp. 213-234.

52. Parsons, Talcott, (1951) The Social System, Routledge,
575pp.

53. Pilisuk, Mark and Rapoport, Anatol., (1963) A Non Zero
Sum Game Model of Some Disarmament Problems.   Peace
Research Society International papers, Volume 1, pp. 57-78.

54. Pool, I. and Kessler, A., (1965) The Kaiser, the Tsar and
the Computer: Information Processing in a Crisis .
American Behavioural Scientist 8, pp. 31-38.

55. Rapoport, Anatol, (1960) Fights, Games and Debates.
Michigan University Press, 400 pp.

56. Rapoport, Anatol and Urwant, Carol, (1962) Experimental
Games: A Review in Behavioural Science, Vol 7, pp. 1-37.

57. Rapoport, Anatol., (1964) Strategy and Conscience. Harper
and Row, 323 pp.

58. Rapoport, Anatol. (1966) Two Person Game Theory, The
Essential Ideas. Michigan University Press, 229pp.

59. Rapoport, Anatol. Formal Games as Probing Tools for
     Investigating Behaviour Motivated by Trust and Suspicion.
     Journal of Conflict Resolution, Volume VII, Number 3,
     pp. 570-579.

60. Rapoport, Anatol., (1967) Use and Misuse of Game Theory.
     Scientific American Vol 217, pp. 50-56.

61. Rapoport, Anatol, (1967) Games Which Simulate Deterrence
     and Disarmament.   Peace Research Reviews, Volume 1,
     Number 1, August 1967.

62. Rapoport, Anatol. (1970) N. Person Game Theory: Concepts
     and Applications. Michigan University Press, 321 pp.

63. Rapoport, Anatol (Editor), (1974) Game Theory as a Theory
     of Conflict Resolution. Dordecht Reidal, 283 pp.

64. Rapoport, Anatol and Chammah. Prisonners Dilemma: A
     Study in Conflict and Co-operation, Michigan University
     Press, 258pp.

65. Rapoport, Anatol and Guyer., (1966) A Taxonomy of 2x2
     Games.   Peace Research Society, International Papers.
     Vol VI, pp. 11-26.

66. Raser, John and Crow, W., (1969) A Simulation Study of
     Deterrence Theories. In, Pruitt, D and Snyder, R., eds.,
     Theory and Research on the Causes of War, Prentice Hall.

67. Reinton, Per Clav, (1967) International Structure and
     International Integration. The Case of Latin America.
     Journal of Peace Research, Number 4, pp. 334-366.

68. Richardson Lewis Fry., (1960) Arms and Insecurity,
     Boxwood Press, Chicago, Quadrangle Books.

69. Richardson, Lewis Fry, (1960) Statistics of Deadly
     Quarrels. Boxwood Press, Chicago, Quadrangle Books.

70. Robinson, Anderson, Hermann and Snyder (1969) Teaching
     with Inter Nation Simulation and Case Studies. In, D.G. Pruit
     and R. Snyder Editors, Theory and Research on the Causes of
     War, Prentice Hall.

71. Rummel, Rudolph, (1966) Some Dimensions in the Foreign
     Behaviour of Nations. Journal of Peace Research, pp. 201-
     224.

72. Schelling, Thomas, (1960) The Strategy of Conflict.
     Harvard University Press,

73. Schelling, Thomas. Bargaining, Communication and Limited
     War . Journal of Conflict Resolution, Number 1, pp. 19-36.

74. Schelling, Thomas, The Strategy of Conflict: Prospects
     for a Reorientation of Game Theory.   Journal of Conflict
     Resolution, Volume II, Number 3, pp. 203-264.

75. Schmidt, Herman, (1968) Politics and Peace Research.
     Journal of Peace Research, Number 3, pp. 235-243.

76. Schofield, Norman., (1975) The Theory of Games.
     European Consortium for Political Research, 103pp.

77. Shubik, Martin, (1954) Readings in Game Theory and
     Political Behaviour, Doubleday, 74 pp.

78. Shubik, Martin, (Editor), (1964) Game Theory and
     Related Approaches to Social Behaviour. Wiley.

79. Shubik, Martin. Some Reflections on the Design of Game
     Theoretic Models for the Study of Negotiation and Threats.
     Journal of Conflict Resolution, Volume VII, Number 1, ,
     pp. 1-12.

80. Shubik, Martin, (1975) Games for Society, Business and
     War: Towards a Theory of Gaming. Elsevier, 371pp.

81. Singer, J. David. ed., (1968) Quantitative International
     Politics: Insights and Evidence, International Yearbook
     of Political Behaviour Research, Vol VI, New York, Free
     Press.

82. Singer, J. David, (1979) Explaining War, Sage, 328 pp.

83. Sisson, Roger and Ackoff, Russell, (1966) Toward a
     Theory of the Dynamics of Conflict.   Peace Research
     Society International Papers, Volume V, pp. 183-197.

84. Sorokin, Pitrim., (1937) Social and Cultural Dynamics
     (4 Volumes). New York, American Book.

85. Tanter, Raymond, (1966) Dimensions of Conflict Behaviour
     Within and Between Nations. 1958-60 . Journal of Conflict
     Resolution, pp. 41-64.

86. Von Neumann, John and Morgenstern, Oskar, (1953) Theory
     of Games and Economic Behaviour, 3rd Edition, Princeton
     University Press.

87. Wiberg, Haken, (1968) Social Position and Peace
     Philosophy. Journal of Peace Research, Number 3, pp. 235-
     243.

88. Williams, J.D., (1966) The Compleat Strategyst. McGraw
     Hill.

89. Wright, Steve, (1979-80) A Multivariate Time Series
     Analysis of the Northern Irish Conflict. Papers, Peace
     Research Society, Vol. 29.

90. Wright, Q., (1941) A Study of War. Chicago, University
     of Chicago Press.

91. Young, Oran (Editor), (1975) Bargaining: Formal Theories
     of Negotiation, Illinois University Press, 412 pp.

THE WEIBULL DISTRIBUTION IN THE STUDY OF INTERNATIONAL CONFLICT*

Ib Petersen
(Institute of Political Studies, University of Copenhagen)

1. INTRODUCTION

Models of stochastic social processes

In a seminal report from the Copenhagen School of Economics
and Social Science 1971, two Americans and a Norwegian
published a model of organizational anarchy which they called
"A Garbage Can Model of Organizational Choice". These two
American fellows were Michael D. Cohen from University of
California and James G. March from Stanford Unviersity. Their
Norwegian counterpart was Johan P. Olsen from the University
of Bergen. The model constituted a break with the past in
that they stressed the stochastic character of collective
decision-making instead of relying on conventional rational
choice models.

A "Garbage-can model" is a theoretical model for decision-
making in so-called "organizational anarchies" in situations
of ambiguity. The model views organizational decision-making
as a dynamic process, a continuous flow, in which situations
are characterized by ambiguity of intentions and comprehension,
where available means or technology are not readily applicable
and where the attention of the decision-maker is changing all
the time. The various elements of decision-making, i.e. the
flow of opportunities for decision, the flow of problems to
be solved, the flow of possible solutions and the flow of
participants are seen as largely independent of one another and
capable of being combined in a number of ways.

* This paper is part of a study which has received financial
assistance from The Danish Ministry of Education, The British
Council and The Institute of Political Studies covering travel
expenses in Great Britain.

The access structure, the ways decisions are made, the
energy expended by participants and the attention paid by
them to problems determine the way the elements are combined.
This structure may vary from complete randomness to a more
ordered one, but it would always contain a large element of
randomness. The result of such a decision-making process would
not necessarily be a combination of a problem with a solution.
It could as well result in a flight from the problem or a
postponement of it.

By contrast, the rational decision-making model assumes no
ambiguity in preferences or the definition of the situation. It
assumes that a number of means are at hand, among which the
decision-maker could choose the most efficient one to obtain a
solution to the problem. Hierarchial or bureaucratic rules
would define the combination of the elements of the decision
and the responsible decision-maker would not shy away from a
decision. The decision-maker is also assumed to pay attention
to the problem in a structured and sustained way.

Another stochastic model was formulated in 1970 by the
American sociologist Harrison White from M.I.T. He sketched a
stochastic model for social interaction using as an analogy
nodes in a communication network. Different roles correspond
to different networks. To cite from his paper presented at
the IPSA conference in Munich:

"Role frames correspond to analytic networks, and what
ties them together to become concrete social structure is the
use of real time by concrete persons who appear in roles in
different networks".

In order to understand the priorities that determined the
flow of messages through the networks, Harrison White pointed
out, we could not just measure the time a person spent on
different activities because he would change back and forth
between different networks in a stochastic way.

The problem of empirical description

The acceptance of the stochastic nature of political
processes has been very slow. Recently the application of
rational choice models has once more become a fashion.

One of the reasons for this may be the virtual impossibility
of describing stochastic processes like the ones mentioned
except by simulation. Cohen, March and Olson recognize in
their original paper the immense complexity of interaction in
a system with even a few specifications. They write:

"Though the specifications are quite simple their
interaction is extremely complex, so that
investigation of the probable behaviour of a
system fully characterized by the garbage can
process and our specifications requires
computer simulation. We acknowledge immediately
that no real system can be fully characterized
in this way." [1].

Harrison White relied on computer simulations of
communication networks carried out by the engineer Leonard
Kleinrock [3]. The applications of such models have not been
very common in political science and some of the applications
have been considered of little use.

To cite an instructive case: In 1972, David Singer from
Michigan University used the exponential distribution to
describe the time periods between the outbreaks of war over
150 years [7]. In his book, this observation is left without
further ado: He goes on to study other variables such as
"the annual amount of war". As judged by the intentions of
Singer's project, to find the causes of war, the observation
was of little value. It says that the probability of war is
independent of the time passed since the last war broke out.
The rather gloomy picture is one of an equilibrium in the process
of onset of war over the 150 years. No variation exists that
could help him form a hypothesis about some cure for the
disease. In fact this information must have seemed detrimental
to all further study of the process of onset of war. The
application of a stochastic model was not found helpful. I
think that my own experience has told me that stochastic models
in political science are poorly understood, considered of
marginal importance and met with outright scepticism. They are
not seen as indispensable tools for the description of
political and administrative phenomena that do not conform to
conventional models of rational behaviour and for the
development of theories of such phenomena.

In fact stochastic processes may be much broader phenomena
than normally recognized, as indicated by Harrison White's
reasoning. In 1971 I discussed his model with Harrison White at
M.I.T. He agreed that it would perhaps be possible to model
the traffic of messages in the networks by measuring the
interarrival time between such messages. If this time-series
would fit a stochastic model it might be possible to find
changes in parameters that reflected changing priorities.

At that time I had already carried out experiments with the
Weibull distribution for some years.

The Weibull distribution

I became interested in stochastic models twenty years ago
when I joined a peace-research group in Copenhagen associated
with the Nordic Summer-University. Here, researchers from
many faculties joined hands in a common endeavour to highlight
the causes of conflict. In this group, psychologist Iven
Reventlow had shown that the behaviour of fish could be
described by means of stochastic models. After extensive
experiments he and his statistical adviser Professor Georg
Rasch had decided that the exponential distribution had to be
rejected whereas the Weibull distribution fitted the
experimental data well.

The distribution is named after Waloddi Weibull, a Swedish
physicist who used it in 1939 to represent the distribution of
the breaking strength of materials. A more widely available
account by the same author was published in 1951. In Russian
literature it is sometimes called the Weibull-Gnedenko
distribution. The use of the distribution has been mainly
within the fields of reliability and control work. From the
early seventies it has become part of the standard repertoire
in the bio-medical sciences under the general heading of
survivor-distributions.

One of the standard descriptions of the Weibull distribution
and a lengthy discussion of different methods of estimation
is found in [2], together with a comprehensive list of
literature. A short introduction to the Weibull distribution
could also be found in my book: Dynamic Laws of International
Political Systems, [5], p. 17-22. (Adapted from Keiding and
Vaeth: >>Forelaesningsnoter i statistisk analyse af
overlevelsesdata<<, Copenhagen 1978).

The renewal process with Weibull-distributed interarrival
times is a simple extension of the Poisson process which can
be specified as:

$$P[T \leqslant t, \lambda] = 1-e^{-\lambda t}$$

A new parameter α is added as an exponent to t, to yield the
expression:

$$P[T \leqslant t, \lambda, \alpha] = 1-e^{-\lambda t^{\alpha}}$$

This is an expression of the probability for a change in the
situation in the time period t, the Weibull cumulative
distribution function.

P = Symbol of probability
T = The stochastic variable "Time".
t = An observation of the stochastic variable (T).
α = The shape parameter
λ = The scale parameter
e = The log base of the natural logarithm(ln).

The conditional probability for a change in the situation observed in the time-span (t, t + Δt), subject to the condition that no change has occurred up to time t is:

$$\lambda(t) = \lambda \alpha t^{\alpha-1}$$

This function, the intensity function, has interesting qualities as a model for stochastic political processes.

If alpha = 1, the intensity = λ. This is the exponential distribution. The intensity of the occurrence of the phenomenon in question is unrelated to the time passed since the last change took place.

If alpha < 1, the probability of a change will decrease with the time passed since the last change.

If alpha > 1, the probability of a change will increase with the time passed since the last change took place.

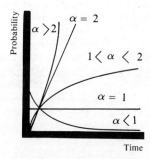

Fig. 1   Intensity curves for varying values of the shape parameter (scale-parameter constant).

Within the context of a renewal process this means that we could model processes of equilibrium, growth and decay.

If we allow for changes in the values of the parameters at definite time points we would be able to model different states of the same system. If over time these states repeat themselves in a certain order we would be able to model a cyclical process. If a disturbing factor has changed the

parameters of the system and the system returns to normal
after the disappearance of this factor we would have observed
a case of "ultrastability", a concept used in system theory.
A number of case-studies of various stochastic processes
carried out over the years have been published in [4] and [5].
Before looking at some case studies, it is appropriate to make
some comments on the methods underlying the studies.

A note on data and method

These stochastic processes are *point processes in
continuous time*. The events take place instantaneously or
within a very small time-interval as compared with the
interarrival time, e.g. the signing of a particular order,
the publication of a note, a declaration by a spokesman, a
conference lasting up to half an hour etc. etc. In one
instance (the series on international negotiation) we have
a semi-markov process. Time periods between summit-meetings
constituted one series and the duration of those meetings
another series.

The *real-time series* utilized in the study have been
constituted by the use of diaries (the Fuehrer conferences),
white books (the Sino-Indian case), data-sets collected by
researchers (international war by Singer and Small, summit-
meetings by Johan Galtung and the "World Event Interaction
Study" by Charles McClelland). One data-set concerning the
Sino-Soviet border conflict has been collected by myself. Raw
data, computations, tests, methods and interpretations can be
found in [5]. The book also contains a detailed account of
the way long real time series covering 10-150 years have been
subdivided into smaller series with significantly differing
shape or scale parameters. A few words on this are in place,
as textbooks normally do not take this possibility into
account. The application of the Weibull-distribution has
mostly been to samples of patients treated by different
methods in order to find the hazard rate or the survivor
distribution of these samples.

The most obvious way of subdividing is to use a method close
to the medical one, namely to subdivide according to a theory
of the importance of a particular event or kind of events. We
could then see whether the occurrence of this event (in
medical science, the treatment) meant that we ended up with
two distributions with significantly differing shape or scale
parameters.

If the history of a given system is well known, it is not
too difficult to formulate ad hoc theories and hypotheses about
changes in the system that would affect the stochastic process.

These hypotheses could then be tested against the distributions
resulting from the subdivision of a long series. On the other
hand, if you only rely on this procedure you may be liable to
commit a type two error. A way to evade this would be to scan
the numerical material by some method in order to see whether
such scanning would result in further hypotheses about
changes in the stochastic process. These hypotheses should
then be tested against the historical evidence available.
(This last procedure should also be played against Monte-Carlo
tests in order to gauge the possibility of forming false
hypotheses.)

Here, the stochastic processes are *renewal-processes*. This
means that the intensity curves displayed below are related
to time-periods in between events. The intensity function is
uniform for a given distribution (series of events). It goes,
however, to zero when an event takes place. After the event
the probability curve starts to rise again.

In the following paragraphs I will give a summmary of
results from the application of the Weibull distribution to
stochastic political processes related to international
conflict.

2. CASE STUDIES

Political-Administrative processes in Nazi Germany 1939-41

The first series of experiments was with political-
administrative processes in Nazi Germany 1939-41. The time
periods between Hitler's meetings with the commanders of the
three armed services and with foreign diplomats and politicians
were measured. These four real-time series were tested for
randomness and shown to fit the Weibull distribution. Each of
them consisted of a number of Weibull distributions with the
same shape parameter (alpha) but changing scale parameters
(Lambda, Tau).

We had thus been able to describe changing states of a
system. It could be shown that the changes in the values of
the parameters for the three series of meetings with
commanders was alternating between two states, i.e. we had
shown the existence of semi-cyclical processes/ultra stability.

The shape parameters in all three cases were greater than
1, which meant that we had processes of growth. The
probability for meetings was growing with the time elapsed
since the last meeting. (This I interpreted as signifying the
high pressure in the Nazi-political administrative system to
regain the strategic initiative lost to England by the untimely

outbreak of war in 1939, that found Germany without an
adequate fleet and long range bombers to fight a war with
England).

The changes of the scale parameters were connected with
the different campaigns. In campaigns where a particular arm
was involved the average waiting time between meetings was
small. Between campaigns long average waiting time designated
a more normal state of affairs.

The meetings with foreign diplomats and politicians, on
the other hand, reflected changing diplomatic offensives to
bolster the German position. In this series we notice three
levels of activity insteady of just two. A very slow process
seems to reflect a process bordering on non-use of the
diplomatic weapon. (Data and values of parameters, including
interpretations of the various phases of activity can be
found in [5].)

The Sino-Indian Border conflict 1959-64

The Sino-Indian border conflict represented an opportunity
to model interactive behaviour in an international conflict.
The result is highly interesting in terms of intensities. Time
periods between aggressive verbal behaviour (notes) were far
from being identical for the two parties to the conflict.
Parameters differed for their respective distributions.
However, if we look into graphs of the intensity-functions we
observe a very pronounced symmetry for the periods preceding
and following the pre-crisis and crisis periods in 1962.
(See Figure 2). We are also able to observe a conflict spiral
that ends up in a short war.

In the precrisis situation, India raises her intensity
curve to a level higher than the corresponding Chinese curve.
It is quite clear that at this juncture India wants, by its
forward policy, to press China out of those areas of the
borderland already occupied by China. India deters China from
challenging its army patrols with might or the threat of
outright war.

This in fact worked for a few months, as China feared an
invasion from Taiwan and could not cope with both Taiwan and
India at the same time. As the threat from Taiwan receded in
June 1962 however, China immediately reversed the situation
by raising its own intensity curve to a much higher level than
India. This time China threatens India. In fact for a short
time Nehru tries to pull out, but public opinion in India has
been raised to such heights that it is impossible to back down.
This means that China's attempt to threaten Indian fails and
Nehru desperately tries to deter China again by raising the
intensity curve to extreme heights.

The trick does not work. China is not deterred, and within
a very short time the Chinese army annihilated the Indian army
in the Himalayas. (On this study of Sino-Indian conflict cf
[5], p. 71 ff).

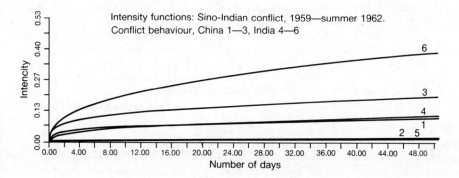

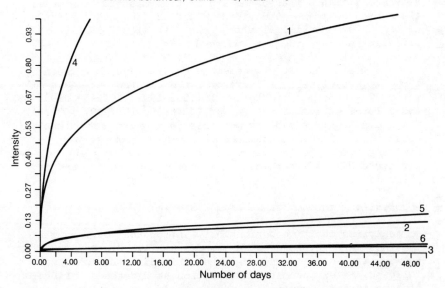

Fig. 2

The pattern of conflict behaviour: The pattern of the paired
conflict behaviour of China and India is quite interesting to
watch. In the first two phases (1959 summer 1962) the intensity
of the conflict is rather low and very symmetrical. In the
third phase the intensity of the conflict behaviour of India
rises considerably so an outspoken asymmetry is obvious, as
China does not raise its own level of intensity very much. The
fourth phase (1962-64) is characterised by very high conflict
intensities and an even greater asymmetry. (India very high,

function no. 4, China high, function no. 1). In the fifth
phase (after the Chinese attack) we are approaching the former
low level of intensity of the conflict and little asymmetry
(functions 2 and 5). In the last phase, conflict intensity is
low on both sides and the situation very symmetrical. High
intensities of conflict and asymmetrical situations seem likely
to lead to open conflict through a conflict spiral. (India
first raises its level of intensity, China responds in the next
phase with a high level of intensity, which spurs ahead India
to a very high level of conflict intensity. War follows).

War in the international system

Moving from the level of the dyad in the international
system to the level of the total system, I reanalyzed David
Singer's data on the onset of war 1815-1965. I discovered
that even if this fitted the exponential distribution well,
the period could be subdivided. Instead of one distribution a
number of exponential distributions with significantly
different scale parameters emerged.

This meant that it was possible to distinguish between
three levels of war; a level of extremely few outbreaks of
war (1823-1848); a middle level (1866-1911, 1919-39, 1951-65)
and a level of much onset of war (1846-66, 1911-19, 1939-51).
These distinctions made it possible to support a hypothesis
about a causal relationship between international power
structures and the frequency of wars: that changes in the power
structure of the international system cause tension and war.
([5]p. 173).

Negotiations in the international system

Analyzing the time periods between international summit
meetings between representatives of the five Big Powers
between 1941-61, I discovered that this distribution also
followed the exponential distribution as did the distribution
of time periods between the outbreaks of international wars.

As seen from this angle, both war and negotiation are stable
features of the international system. However, this series
could also be subdivided. A number of exponential
distributions with significantly differing scale parameters
again emerged and three levels could be ascertained; a level
of very few summit meetings (1947-53), a normal level of
activity (1941-47, 53-59, 60-61) and a level of many summit
meetings (1959-60). At the same time it could be shown that
the duration of summit meetings could be described as a number
of Weibull distributions representing processes of growth and

decay. In accordance with the garbage-can model, time spent
on summit meetings represents energy expended in the collective
decision-making process. The frequency of meetings also
represents a measure of the energy spent in the process
(Cohen, March, Olsen 1971).

We can distinguish between periods with long meetings
(1945-50, 1953-55) periods with medium long meetings (1941-45,
1958-59) and meetings of short duration (1950-53, 1955-58,
1959-61).

Energy spent in the collective decision-making process thus
varied over time as follows:

1941-45: Medium frequency, medium duration
1945-47: Medium frequency, maximum duration.
1947-50: Minimum frequency, maximum duration.
1950-54: Minimum frequency, minimum duration.
1954: Medium frequency, maximum duration.
1955-1958/59: Medium frequency, minimum duration.
1959: Maximum frequency, medium duration.
1960: Maximum frequency, minimum duration.
1960-61: Medium frequency, minimum duration.

According to conventional classification schemes in large
international datasets such as "WEIS" and COPDAB", summit-
meetings represent cooperative acts. On this basis it is
quite obvious that Great Power cooperation was already
diminishing before the outbreak of the Korean war on 25th June
1950. Furthermore, the minimum cooperation period closely
follows the war period.

Maximum cooperation periods closely match periods of
detente:

1945-47 was a period of allied post-war reconstruction,
1954 brought armistice in Vietnam,
1959 was the year of the Camp David meeting between Eisenhower
and Kruschev.
A detailed account of the study can be found in [5]p. 185 ff.

*Verbal conflict behaviour in the relationship between US and USSR**

This interactive conflict system is a much more complicated system than the Sino-Indian conflict system. Over time both parties change not only scale parameters (as was the case with China and India) but also their shape-parameters.

Furthermore, as can be seen from the paired intensity functions shown in Figure 3, asymmetrical relationships are typical. We find only a few approximately symmetrical relationships.

The intensity curves are conflict curves. The level of the USSR curves are generally higher than the US curves: in fact for the first 10 years or so the US keeps a very low profile except for a few crisis situations. This could plausibly be related to the protracted Vietnam war. As we know, Lyndon Johnson acted with restraint in not invading North Vietnam, even if air bombing on a large scale was carried out. We observe USSR react vigorously to the first large scale bombing campaign in the spring 1966 and later to the Christmas raids over Hanoi and Haiphong 1972, preceding the signing of agreements in January 1973. (On the other hand, equally vigorous bombing in 1967 was not reacted to by the USSR!).

---

* The data utilized in this study was made available in part by the Interuniversity Consortium for Political and Social Research. The data were originally collected by Charles A. McClelland, University of Southern California. Neither the original collectors of the data nor the Consortium bear any responsibility for the analyses or interpretations presented here.

The computation of the time-series was done by Magnus Magnussen and most of the work with the subdivision was carried out by Henrik Holtermann who has been associated with the study from the very beginning. He is presently carrying out a special study of the use of the armed forces by both superpowers for political purposes.

Magnus Magnussen has carried out special tests for randomness utilizing a program also geared to find even very complicated associations among categories in the Weis data. The result was negative to a point that M.M. found "surprising".

The final report will be published in 1986 in a book from the Copenhagen Political Studies Press.

THE USSR-US CONFLICT SYSTEM 1966-78.

Thick lines US intensity curves - **Thin lines** USSR intensity curves.

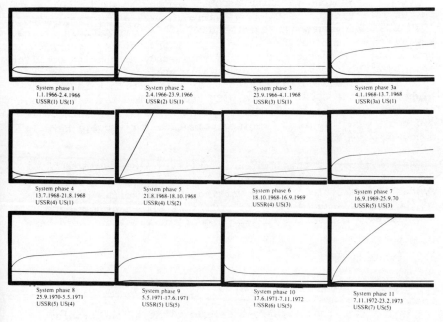

System phase 1	System phase 2	System phase 3	System phase 3a
1.1.1966-2.4.1966	2.4.1966-23.9.1966	23.9.1966-4.1.1968	4.1.1968-13.7.1968
USSR(1) US(1)	USSR(2) US(1)	USSR(3) US(1)	USSR(3a) US(1)

System phase 4	System phase 5	System phase 6	System phase 7
13.7.1968-21.8.1968	21.8.1968-18.10.1968	18.10.1968-16.9.1969	16.9.1969-25.9.70
USSR(4) US(1)	USSR(4) US(2)	USSR(4) US(3)	USSR(5) US(3)

System phase 8	System phase 9	System phase 10	System phase 11
25.9.1970-5.5.1971	5.5.1971-17.6.1971	17.6.1971-7.11.1972	7.11.1972-23.2.1973
USSR(5) US(4)	USSR(5) US(5)	USSR(6) US(5)	USSR(7) US(5)

THE USSR-US CONFLICT SYSTEM 1966-78.

Thick lines US intensity curves - **Thin lines** USSR intensity curves.

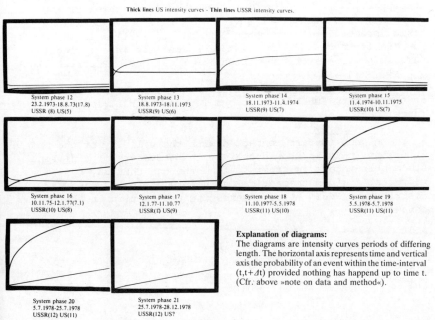

System phase 12	System phase 13	System phase 14	System phase 15
23.2.1973-18.8.73(17.8)	18.8.1973-18.11.1973	18.11.1973-11.4.1974	11.4.1974-10.11.1975
USSR (8) US(5)	USSR(9) US(6)	USSR(9) US(7)	USSR(10) US(7)

System phase 16	System phase 17	System phase 18	System phase 19
10.11.75-12.1.77(7.1)	12.1.77-11.10.77	11.10.1977-5.5.1978	5.5.1978-5.7.1978
USSR(10) US(8)	USSR(11) US(9)	USSR(11) US(10)	USSR(11) US(11)

System phase 20	System phase 21
5.7.1978-25.7.1978	25.7.1978-28.12.1978
USSR(12) US(11)	USSR(12) US?

Explanation of diagrams:
The diagrams are intensity curves periods of differing
length. The horizontal axis represents time and vertical
axis the probability of an event within the time-interval
$(t, t+\Delta t)$ provided nothing has happend up to time t.
(Cfr. above »note on data and method«).

Fig. 3

Strong reactions can also be observed on the US side to the
Soviet intervention in Czechoslovakia in 1968 and in May-June
1978 in connection with Soviet and Cuban intervention in Africa.
Now and then we have what seems to be a prelude to a conflict
spiral but within the confines of this study no genuine conflict
spiral can be found. (Note especially the gradual rise of the
US conflict curve during phases 17-19 ending with a level much
higher than the USSR. If the USSR had accepted the challenge
in phase 20 a genuine conflict spiral would have started. It
did not). The greatest combined potential for trouble here is
found in May-June 1978.

As the study of this system is still in progress, no
detailed interpretation of the development of the conflict
system is available. We are working on time series related
to the use of armed forces for political aims that may be a
corollary to the series on verbal conflict behaviour used here.

3. PROBLEMS

Validity of models

Apart from minor technical problems, one of the problems
with the application of stochastic models is connected with
the way features of the model are used to represent theoretical
concepts. Is it, for instance, reasonable to think of the
intensity as a measure of energy in a collective decision-
making process? (Cohen, March and Olsen, 1971). Is it
reasonable to measure the risk of a conflict in a collective
bargaining process by the intensity function? (Petersen 1980).

If so, what is the possible meaning of other features such
as the density function, the cumulative distribution
function etc. etc.?

These problems are connected with the validity of the model.
It follows that we should analyze the structure of different
stochastic models in order to find the one that may best
represent a given theoretical structure. (The Gamma-
distribution would be an interesting proposition in this
connection).

Theories of underlying processes

The theoretical understanding of stochastic processes in political conflict should be broadened. As it is we have very little theorizing of this kind, apart from the garbage-can model and occasional papers on organizational anarchy (see also Yolles [9], this volume).

Reliability

A thorough discussion of the techniques used should result in the approval of a set of standard procedures to ensure reliability, e.g. of the way real-time series should be divided into several significantly differing distributions.

REFERENCES

1. Cohen, M.D., March, J.G. and Olsen, J.P.,   (1972)   A Garbage Can Model of Organizational Choice.   Copenhagen School of Economics 1971.   (Also published in Administrative Science Quarterly, Vol. 17, pp. 1-25).

2. Johnson, N.L. and Kotz, S., (1970)   Continuous Univariate Distributions, Vol. 1, Boston.

3. Kleinrock, L., (1964)   Communication Nets, Stochastic Message Flow and Delay.   New York.

4. Petersen, Ib.D., (1979)   Kybernetiske systemers udviklingslove, Copenhagen.

5. Petersen, Ib.D., (1980)   The Dynamic Laws of International Political Systems.   Copenhagen.

6. Reventlow, I., (1969)   Studier over Komplicerede psykobiologiske faenomener.   Copenhagen.

7. Singer, D. and Small, M., (1972)   The Wages of War 1816-1965.   A Statistical Handbook.   New York.

8. White, H., (1973)   Paper presented at the IPSA conference in Munich 1970, published in a revised edition in: Alker, Deutsch and Stoetzel: Mathematical Approaches to Politics. Elsevier.

9. Yolles, M., (1984)   Modelling Conflict with Weibull-games, IMA conference, Cambridge, December 1984.

10. Keiding, N. and Vaeth, J., (1978)   Forelaesningsnoter i statistisk analyse af overlevelsedata, Copenhagen.

SOME IDEAS TO HELP STOP THE ARMS RACE

S.H. Salter
(University of Edinburgh)

1.   INTRODUCTION

Despite the strong wishes of a large majority of the
world's people, progress in reaching agreement on the reduction
of nuclear weapons has been very slight. Despite the work of
our cleverest brains and the spending of ever-increasing amounts
of money, security seems ever more remote and the issues appear
ever more complicated. The arguments grow into head-spinning
iterations of "we think that they think that we think that
they think". While some experts are close to despair others
would have us believe that the problems are too difficult for
us outsiders to understand and that solutions have to be left
to them - the successors of the people who began it all.

Perhaps it might help to study a simpler way to get to a
very stupid position by a series of apparently sensible steps.
The analogy [7] is due to Samuel Gorovitz, Professor of
Philosophy at the University of Maryland.

Gorovitz describes an auction. As in a normal auction the
goods are sold to the highest bidder. The difference is that
the second-highest bidder also has to pay his bid but gets
nothing. Let us suppose that the article for sale is a dollar
bill and that the auctioneer invites an opening bid of ten
cents. The profit margin appears very attractive and plenty
of people will be found willing to make a small bid for such a
handsome return. Unfortunately there will be others willing to
raise the bid to fifteen cents. While it is splendid to buy
a dollar bill for only ten cents it is not good to pay ten
cents and get nothing. The sensible step for the opening
bidder is clearly to raise the bid to twenty cents.

The two bidders are now locked into the fiendish trap devised by the rules of the auction. As bidding proceeds, their possible profit steadily declines while the penalty for coming second steadily rises. A smile appears on the face of the auctioneer as the bids pass fifty cents. The smile gets wider as each victim tries to force the other into second place and the bids get bigger. In psychological experiments it has been shown that people will bid up to five times the value of the article which they were originally tempted to buy for a tenth of its value - a fifty-to-one change. The auctioneer wins ten times his risk capital.

We should note that it is only the rivalry between the bidders that gives the auctioneer his chance. If they could have been persuaded into cooperating by making a joint single ten cent bid they could have shared the dollar and put the auctioneer out of business. There is no record of experimental subjects falling for the trick twice but presumably a sufficiently dogmatic insistence on out-bidding a competitor would lead to this result.

Despite the simplicity of its rule, the Gorovitz auction is an acutely accurate model of the progress of an arms race. In 1945 it seemed as though nuclear weapons would give everlasting peace. There need be no more military service or dead young men. The initial bid was low. A very few bombs would do.

But neither super-power could allow the other to make the higher bid. Each increase had to be matched. Inevitably, as time has gone on, the size of the bids has risen. We keep hoping that each new one will provide the longed-for security but each in turn proves inadequate. It is most interesting to retrieve from the archives the proposals for yesterday's weapons, now claimed to be insufficient, and to predict the inadequacies of the ever more expensive weapons proposed for tomorrow.

The solution of problems can often be indicated by careful identification of the difficulties. Let us therefore write down a list of the reasons why progress in nuclear arms reduction has been so slight.

1. Mutual suspicion

2. The excessive size of proposed reduction steps

3. Loopholes generated by previous agreements

4. The difficulties of weapon comparison

5. The perception of unequal starting points

6. Anxieties about verification

7. Retentionism - that is, the determination of an
    industrial or military group to continue armament
    expansion.

It turns out that some of these difficulties can be
minimised and others actually turned to advantage by an
unconventional approach. Let us consider the difficulties
in more detail.

2. OUTLINE OF THE PROBLEMS

Suspicion

Mutual suspicion between the two sides is very deep. The
original reasons for it have been overlaid by the nuclear arms
race itself. Each side used to make nuclear weapons because it
hated the other's social system. Now it makes nuclear weapons
because the other makes nuclear weapons. Any suggestion made
by one side is rejected instantly. The fact that Side A
proposes something is taken as quite sufficient evidence that
it must be to the disadvantage of Side B even if no disadvantage
is apparent.

I conclude that a successful disarmament process should be
totally symmetrical and that the need for dialogue should be
minimised.

Excessive step-size

Proposals for sweeping arms reductions have a dramatic
quality which makes them attractive to voters. But it is
reasonable to expect that the anxieties of accepting a
reduction will be proportional to its magnitude. It should
be easier to take the risk of a very small step since there is
time to consider the position after it.

I conclude that a series of very small reductions is more
likely to be accepted than a 'zero option' plan. Tension
would be reduced if the rate of increase were to slow. A total
freeze would be even better. But an actual reduction - no
matter how small - would produce a relaxation of tension much
greater than its military significance.

Loopholes

Foolish classification formulae can produce gross
abnormalities in the design of racing cars and yachts.
Similarly, disarmament agreements which prohibit particular

types of weapon will lead to frantic developments in the areas
which escape the ban. If there are limits on the number of
launchers we get multiple warheads. If there are limits on
throw-weight we get miniaturisation, high-yield explosives
and terminal guidance. If long-range strategic weapons are
stopped then there will be forward-based, short-range tactical
ones. If quantity is curtailed, quality will improve. It is
therefore not clear that the results of over-specified arms
restraints leave the participants any better off.

I conclude that weapon definition in a successful
disarmament system should be as plain and broad as possible.

Weapon Comparison

Endless measurement difficulties bedevil the comparison of
weapons. There is the destructive power of the warheads and
the payload of the launchers. There is the time required for
launch preparation and the flight time to the target. There
is the accuracy of the guidance systems and the sophistication
of electronic counter-measures. There is the flexibility of
target alteration, and the mobility and detectability of
carrier vehicles. There is the hardness of the silos and the
bandwidth and immunity of communication networks. There is
the radar cross-section of aircraft and missiles. There is
the sonar signature of submarines. Every judgement will change
in response to technical advances on either side. Comparisons
would be difficult enough in calm debate among the officials
of one side equipped with accurate knowledge of their own
weapons. But when the uncertainty of information gained by
espionage is combined with hostile international relationships
the problem becomes quite intractable.

I therefore take it as axiomatic that disarmament proposals
which rely on exact weapon comparisons will fail. As we shall
see, however, it is possible to turn the difficulty to
advantage and design a procedure which works better because
of differences of opinion.

Unequal starting points

The two sides claim that disarmament might be possible if
only their starting points could be equalised. By concentrating
on particular categories of weapons deployed in particular
areas both sides can produce arguments to suggest that the
other is building a dangerous advantage.

Let us examine every possible route to a reduction as seen
through the eyes of two opponents. To avoid any negative
associations the sides are called Dot and Dash. In the
following graphs the horizontal axis is time, while the
vertical axis shows perceived lethality measured by some
complicated index of quantity and quality of weapons,
associated launchers, command networks etc. Reductions can
be arithmetically equal or proportional to original holdings.
Initial steps either upward or downward can be made to
equalise starting points. The changes made by one side are,
of course, seen as different magnitudes by the other.

Dot's view   Dash's view

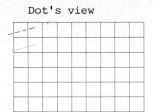   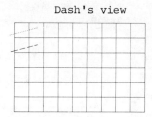

The claimed perceptions of the starting positions will be
like this.

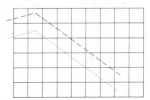   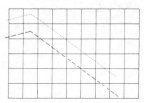

If it were possible for an independent measurement system to
suggest arithmetically equal reductions, then the two curves
would descend along lines of equal slope. Each side
(considering itself to be the initially inferior group) feels
that these arithmetically equal reductions will steadily worsen
the ratio from their point of view, leading to a dangerous
instability when one side has nearly nothing and the other is
superior by the original difference.

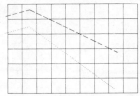   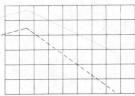

However, because of their mutual suspicion, each side feels that
specific reductions suggested by the other would produce results
more like this, leading to an even wider gap than the
independently-determined arithmetically-equal reduction.

Dot's view   Dash's view

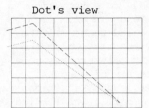   

To overcome this problem we might consider using reductions
proportional to the original holdings. The slopes suggested
by each side would then be like this. The arithmetical gap
between the two steadily closes while the initial ratios
are preserved.

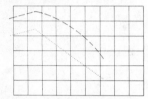   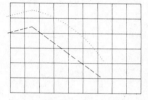

Each side expects, however, that the other will get rid of its
inferior weapons first so that there will again be a dangerous
widening of the gap.

Attempts can be made to equalise the starting position in two
ways. Dot suggests that Dash should first of all give up
such-and-such a portion. If equality could be achieved,
arithmetically equal and proportional reductions would
thereafter be identical.

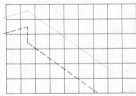

But to Dash, Dot's proposal looks as if it will widen the
initial gap.

Dot's view   Dash's view

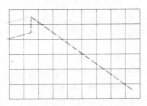

Dash therefore suggests its own holding should first be
increased to match Dot's, from which point reductions can
proceed.

To Dot the scheme would look like this.

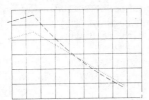   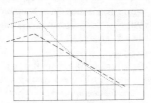

Clearly we have to design a mechanism in which Dash's
reductions look bigger to Dot than they do to Dash, while
Dot's reductions look bigger to Dash than they do to Dot.
The disarmament curves must be seen from both sides to converge
and perhaps eventually to cross. At each stage the feeling of
security must be greater than at the stage before.

3. A POSSIBLE SOLUTION

   A Childhood Parallel

   Given that a child is not suffering from malnutrition, its
absolute happiness is not much affected by whether or not
there is cake for tea. But children are acutely sensitive to
relative benefits. If there is cake for tea then it is
distressing to feel that somebody else is getting more.
Children have devised an elegant solution to the difficulty,
based on the "I cut, you choose" rule. (Readers are advised
to be choosers rather than cutters. A chooser can detect any
minor imprecision in the cut and turn it to slight advantage.

Cutters try to minimise this by being as accurate as possible.)

Mathematicians like cake as much as anyone else and have found rules to share cakes among more than two people [5] [8]. They have also been able to prove the intriguing fact that if there are irregularities in the cake, such as more icing in one place or more cherries in another, and if the opinions of the sharers about the relative values of icing and cherry are different, then all the parties to the division can feel sure that they have come off best. This assurance is exactly what we need for nuclear disarmament.

A cake-sharing disarmament agreement would work as follows. For the first step each side writes out a list of all the weapons in its own nuclear inventory. Against each item it puts a number which I shall call a *military value percentage*. This number represents the military usefulness of the item as seen by its owner. For example, if one side possesses 25,000 warheads of equal usefulness then the military value percentage of each will be 0.004%. But weapons vary a great deal, and so the entries will range either side of that value. The sum of all the military value percentages is made equal to one hundred. The choice of values may involve heated debate between the services of one country but the final result is entirely an internal decision.

The numbers do not represent any sort of absolute value of the weapons because we accepted earlier that no such value could be determined. They are purely the values in the eyes of their owners. It does not matter very much if they are a little inaccurate. Indeed, for a number of reasons, either side may choose to distort them. What matters is that neither side should be able to claim that the loss of one small part of its inventory would reduce its security more than the loss of a different small part with an equal military value percentage. If the loss was different then they could blame no one but themselves for getting the numbers wrong.

It would be extraordinary if the feeling of threat induced by every weapon in its potential victim matched the feeling of security that it provided to its owner. For example (avoiding the indelicacy of precise identification), a mobile, forward-based, quick-launch, unhardened missile with accurate terminal-guidance can very easily be mistaken for a first-strike deterrent-destroyer without providing much protection for its land of origin. However, a less accurate submarine-based weapon which is safe from detection and could be used at leisure for retaliation as a second-strike weapon is a splendid protection for its owner without appearing too threatening to the target. When the two sides look at each

other's lists they will therefore see some weapons that have
been given military value percentages lower than their
perceived threat values.

Now for the crucial step. Each side picks weapons to
some agreed very small total military value percentage - such
as 1% - from the list of its opponent and asks that these be
dismantled in return for a similar amount chosen by the other
side from its own list. It will of course have to pretend
to be quite indifferent as to which of its own weapons are
selected because, supposedly, it chose the numbers to make all
possible choices equal.

Now it is quite clear that, because there was a difference
of opinion about values of security and threat, both sides
will feel that they have obtained an advantage. They have
removed the nastiest-looking devices threatening them and
paid for this pleasure with standardised reduction. It must
be the case that the perceived gap between them has narrowed.
Side A will think that its own inventory is now 99% of what it
had originally. But Side A will also think that Side B's new
inventory is less than 99% of Side B's original one by an
amount which depends on the differences between perceived
values of security and threat. The total number of weapons
in the world will have gone down and the process will have
selectively reduced weapons which have a high threat value
while leaving behind those with a high security value. The
side which argues that the other has a larger quantity
of weapons (which means both of them) will have the satisfaction
of knowing that the absolute reduction in its opponent's
armoury will be larger than its own. *Only a side that
secretly knew it had supremacy and was determined to retain
it could have any logical reservations.*

Differences of opinion about weapon valuation have proved
the stumbling block for other negotiation plans. In the
cake-sharing method such differences are used to provide the
incentive. To show that the mathematics are sound I
attach an example of how wealth can apparently be created
by the differences of opinion during the division of an
inheritance.

Three heirs to an estate are to share equally in the
legacy, which consists of a house, a car, a boat, and a
cabin. One approach, of course, would be to sell all
four items and equally divide the proceeds. But the
heirs feel this is unfair, because they themselves
place different monetary values on the objects. Instead,
they say, each heir should receive one-third of what he
perceives to be the worth of the total package. So they
agree to make independent monetary assessments, and this
information is summarized in the following table:

	#1	#2	#3
House:	$32,000	$30,000	$33,000
Car:	$ 3,000	$ 2,000	$ 4,000
Boat:	$ 3,500	$ 4,000	$ 2,000
Cabin:	$ 6,500	$ 6,000	$ 9,000
Total assessment:	$45,000	$42,000	$48,000
One-third share:	$15,000	$14,000	$16,000

Each item is then given to the person who values it the
most. So #3 receives the house, the car, and the cabin,
for a total of $46,000, in his opinion; #2 receives what he
feels is a $4,000 boat; and #1 gets nothing. Now #1 has
a legitimate $15,000 claim against the estate, and #2 has
a $10,000 claim. These claims are paid off by #3, who has
an excess of $30,000 over his fair share. If the extra
$5,000 is divided equally, then each heir will have
received (in goods or money) $1,666.66 more than he
deemed his fair share.

From Kenneth Rebman [8].

    The mix of nuclear weapons should have been carefully
designed to fulfil the defence needs of their owners. The
first reduction may have produced a slight alteration in that
balance, and so at the end of the first step there should be
a pause to allow the military value percentages to be
reconsidered in the light of the remaining inventory and the
new knowledge of the opponent's feelings. Each side must feel

quite confident that it cannot be forced into the position of having an insecure mix of weapons. The chance for adjustment between every round prevents one side from making excessive selections from one section of its opponent's list. If it did so, the military value percentages of the remaining weapons would be raised so that fewer of them would be lost at each reduction. Balance is therefore preserved. Indeed it is possible that, by judicious choice of military value percentages, a side which began the process with a poor balance between its various weapons could move towards a better one. A slow, steady series of microsteps which can be steered by both sides is much less risky than rapid transients.

It is interesting to consider the effects of deliberate distortion of military value percentages. Random deviations on either side of your own best perceived values are of no benefit to you. They give your opponent the chance to make an excess benefit by picking a greater number of under-valued good weapons. Instead you want your opponent to select your over-valued but inferior weapons and leave behind more of your better ones. Distortion should be a legitimate ploy but it is not without risk. One cannot increase the value of one weapon without reducing that of another. If the espionage services of the other side are up to their task and detect the distortion, then the gambit will backfire. I believe that the best policy may be to make one's own lists an accurate reflection of one's own view.

An interesting variant of the scheme has been proposed by Robert Apfel [1] from Yale University. He suggests that Side B should assign military value percentages to the weapons of Side A, who would then choose any items from its own arsenal totalling the agreed disarmament step. This means that the actual items selected are chosen by their owner but that the number of weapons removed is chosen by their potential target. The essential cross-separation of selection and valuation remains. It may be that this variant might make military leaders feel more in control of their remaining weapons. However, political leaders and the civil-defence authorities might feel happier about influencing the weapons aimed at their own populations.

Cake-sharing reductions will provide an interesting channel of communication between the disarming sides. Both the numbers in the military value percentage lists and the types of weapon selected for reduction convey much about intentions. It is possible to write 'friendly' lists of military value percentages which put higher values against your own second-strike deterrent systems and low figures on your first-strike ones. This should entice your opponent to continue

the procedure. It should paradoxically not be possible to
make friendly choices from his list because each choice is,
by definition, equal in the perception of the other side.
The best plan is to select what looks like the best buy to you.
Antagonism is possible when lists are revised, but to give way
to it would be counter-productive: after analysis of his
choices you can try to re-write your own list so that no
particular choice looks specially attractive to him. If he
then makes a random choice, your own security will be
damaged in proportion to your distortion of the items he
chose. While he will perceive no gain, you will perceive
a loss.

List writing

In recent months I have discussed the cake-sharing scheme
with every disarmament expert and official I could find who
was willing to listen. It was a surprise to discover how many
of them expected that getting agreement about military value
percentages between the armed services of one side might prove
a major stumbling block. My innocent views about the
convenience of discussion in the same language, devotion to
duty, military discipline and patriotic loyalty to the Supreme
Commander were badly shaken. It was disturbing to learn that
the generals hate the admirals as much as they hate the enemy
and that this hatred will intensify if military spending is
reduced. We have to devise a scheme for resolving internal
wrangles before tackling the larger task.

Clearly the problem must be soluble: every year we share
the military budget between the services and choose to spend
money on one weapon rather than another. These choices are
declarations of our present perception of the relative values.
One rule for writing the lists would be to base military value
percentages on the original costs of the weapons with allowances
for inflation and obsolescence. Presidential dictat seems
over-tyrannical. A more interesting method uses a technique
which has proved so successful in the thorny disputes between
trade unions and employers that it deserves wider recognition.

The normal procedure is for the unions to ask for a grossly
exaggerated pay rise while the employers respond with a
minuscule offer. The parties eventually converge on a settlement
somewhere between the two initial extremes. Honours are gained
in proportion to how close the final settlement is to one's
starting gambit. The convergence can sometimes be assisted by
a mediator but his presence is not essential.

A less common arrangement requires the presence of a neutral arbitrator whose aim must be the benefit of both sides. The arbitrator is allowed to choose one or other of the proposals in its entirety with absolutely no modifications. Since the more outrageous proposals stand less chance of being chosen by the arbitrator, the initial positions tend to become virtually identical. Honours are gained by making one's proposals appear to be the more reasonable. Mediation and arbitration are quite distinct.

Taking this as a model, each service would write out its list of military value percentages for the inventories of its own and the other services, supporting them with carefully written advocacy. The Supreme Commander, taking whatever advice he pleases, would accept one list for the next round of reductions. There would be no need for the public to know which lists had been rejected. Anxiety about the decision need not be high because of the very small steps involved. The services which had their lists rejected could modify them for the next round, learning by experience. But the lists would be so similar that the selection of a rival one need not be a humiliation.

Freezing and Updating

There can be no doubt that a total freeze on development, testing, production and deployment of nuclear weapons would go a long way towards relaxing tension. Unfortunately the forces opposed to a freeze are very strong. Jobs, promotion, prestige and prodigious amounts of money are at stake. These pressures are enough to corrupt the flow of information and manipulate political processes.

It is at least arguable that some types of new weapon ought to be developed if they provide security at lower levels of threat. The cake-sharing scheme can still operate despite the introduction of new weapons. Side A would propose a new development and suggest a military value percentage for it. Side B would then be entitled to a free selection of equal value. The value eventually chosen would have to be large enough so that Side B would see advantage in choosing its reduction from the older stock rather than immediately selecting the new items. This process would encourage the evolution of weapons with high security but low threat values. One such is described later.

Verification

Anxieties about verification can be used as excuses for rejecting any proposal, no matter how fair. In particular the anxiety about initial starting positions must be overcome.

This problem is, of course, common to all disarmament
processes. But there are two features about cake-sharing
which should minimise these justifiable feelings.

Firstly, it is fair to expect that the larger the
disarmament step, the greater the anxiety about cheating and
the larger the military catastrophe if one is cheated. But
cheating over steps so small as not to affect the overall
balance raises much less concern. If cheating is suspected
then further reductions can be stopped until the matter is
cleared up.

Secondly, we can assume that the probability of detecting
violations will rise with the time available for detection.
There will be more satellite passes and opportunity for
espionage. Intelligence reports can be digested more thoroughly
and correlated more widely. The slowness of the cake-sharing
scheme minimises the very serious difficulties of verification.

The advice to our leaders should be:

'Don't worry about the first few steps because the
reductions are so small.'

'Don't go on to the later steps if you are not satisfied
with the results of the earlier ones.'

'If you suspect cheating give the evidence to your opponent
in private, thereby giving him the chance to correct his
'inadvertent oversight'.'

'Do everything you can to convince your opponent about
your own reductions and goodwill.'

New ideas on verification can help. Some of the money
saved on weapons should be spent on surveillance satellites.
It may be possible to allow partial access to the computer
systems which record the movements and service logs of
individual military units. The data could be handed over after
a delay sufficiently long to remove tactical military risks
and then scrutinised in the manner of tax returns. Our tax
officials can piece together a picture of our accounts which
is generally good enough for their purposes despite our
reluctance to pay tax. There are a number of technically
advanced neutral countries who would be willing to provide
teams for intrusive inspection visits. The technology of
non-intrusive monitoring instruments is well advanced.

History should encourage us: while there have been
accusations of breaches of clauses of existing treaties, only
a few have stood up to close examination. Given the level of

inter-power rhetoric, the observance of agreements - even
those which remain unratified - has been remarkably good.

I am confident that the problems of verification are
soluble. My view must be shared by such well-advised people
as Ronald Reagan and Margaret Thatcher because they would not
otherwise have been sincere in their proposals for the
'Zero Option' plan, which would have depended so heavily on
instant and absolute verification and which would have led to
catastrophic consequences if cheating had occurred.

Verification will be viewed with less anxiety after
the first stages of a cake-sharing reduction. Both sides
can observe the process and draw conclusions long before
their overall capability is threatened.

Secondary Nuclear Powers

The nuclear inventories of the major powers are so much
larger than those of their allies that there is not an
immediate difficulty about the smaller holdings. But as
disarmament continues their significance will grow. It is
quite unfair to expect them to be left out.

A solo power can concentrate the minds of an opposing
bickering alliance by offering to reduce its arms through
sharing them with hitherto non-nuclear partners. Cake-sharing
begins to break down when we have to deal with secret
alliances in which a false ally presses for unwise selections
from the inventory of its supposed opponent. It would seem
prudent to press on with disarmament before proliferation makes
things worse.

The Ultimate Nuclear Weapon

The present absurdly large nuclear inventories are the
result of anxiety felt by each side about the effectiveness
of the weapons they already own. Orders to launch might not
get through in the very short time available. Weapons might
be destroyed before launch or intercepted in flight. Guidance
might be inaccurate and warheads might fail to detonate. The
targets might be harder than expected and enemy civil defence
precautions might reduce the effectiveness of hits. These
fears multiply and re-multiply the numbers. Weapons aimed at
other weapons only breed further ones.

All these fears would instantly be removed if there was
certainty that weapons would get through and would work. If
this was the case, the necessary numbers would be much smaller.
While there is room for discussion I suggest that a number of

warheads greater then ten but less than one hundred would be
adequate to provide all the 'benefits' claimed by advocates
for nuclear technology provided that penetration was absolutely
guaranteed.

This point has led to an intriguing suggestion made
independently by Leo Szilard [10] and Richard Garwin [6]* for
the 'overt emplaced weapon': the best way to be sure that your
weapons get through would be to have them installed at their
target.

This idea rouses instantaneous antagonism whenever it is
suggested, but the rationalisation of this antagonism can
teach us much about our inner feelings. In the Szilard scheme
pairs of cities were twinned with one another, and randomly
selected citizens served a short term of duty manning silos
in the opposite city. A suitably large quorum would be able
to detonate their bomb with themselves beside it but would only
have the motive to do so if their own families had been
destroyed. Self-destruction is perhaps the least dishonourable
way in which a nuclear weapon could be used - better by far
than any cowardly refuge in a hardened bunker. Anxieties
about living in one of the selected cities would be reduced
both by the low probability of the use of the bombs and by
the very attractive subsidies paid to residents. A small
fraction of the savings from the military budget could make
life in the selected cities very comfortable indeed.

Garwin's proposal replaced the silo crew with a
high-technology communication system which detonated the bomb
if an encrypted sequence of non-detonation signals was
interrupted for a predetermined length of time. He
characteristically designed safety mechanisms to the level
of those now used to authorise nuclear attacks and validate
their proper authorisation.

I ask the reader to defer for a moment the instant
revulsion against the prospect of seeing the entrance to a
silo outside his city hall. There can be no logical
difference between the actual presence of a weapon and the
existence of a sufficiently large number of distant ones with
your map reference programmed into their inertial guidance
computers. It is more uncomfortable because you are less able
to suppress an important reality underlying your present way
of life.

----

*Szilard is credited with the conception of chain reaction.
Garwin did the engineering for the first transportable fusion
warheads.

Let us examine the advantages of emplaced weapons

1. The number of weapons would fall from 50,000 to perhaps less than 200 so that the chances of accident are reduced.

2. The time for decision-making is greatly extended. The pressures for launch-on-warning are removed.

3. All the hardware is securely ground-based so that reliability should be enhanced.

4. The holdings of each side can be exactly balanced and gradual reductions can eventually be achieved.

5. In the last resort a level of retaliation short of that required to induce a nuclear winter is assured.

I recognise that there is very little prospect of getting acceptance of the overt emplaced weapon. Nuclear decisions are based on primitive emotional reactions by people who are non-rational enough to fall for the Gorovitz auction time after time.

Lurking behind the overt emplaced weapon is its sinister covert counterpart which can bypass the most expensive 'star-wars' defence system and leave its victims uncertain as to where their massive retaliation should be directed.

4. CONCLUSIONS

1. We have to convince ourselves that arms reduction is possible.

2. Many small steps will be easier and safer to make than a few big ones.

3. The cake-sharing scheme thrives on differences of opinion about security and threat.

4. Cake-sharing selectively reduces weapons with high threat values and leaves behind those with high security ones.

5. Both sides can feel sure that they are gaining at every stage. They will feel that the other side is reducing arms at a rate which is greater than would be the case for equal proportional reductions.

6. It is possible to devise rules to overcome interservice rivalry.

7. Cake-sharing can operate despite the introduction of new weapons. Indeed it can benefit from the introduction of 'nice' new ones.

8. Verification anxieties, though still present, are less acute than for zero option agreements.

9. Only people who secretly know that their side has nuclear supremacy and are determined to retain it can have logical reservations about cake-sharing reductions.

10. There would be several advantages if missile-delivering systems were to be replaced by a much smaller number of overt emplaced weapons.

We should press the leaders of the super-powers to answer this question: 'Despite your own feelings about the malevolence of your opponents, how big a cake-sharing slice would you risk in order to demonstrate your own goodwill?'

ACKNOWLEDGMENTS

I am deeply indebted to Richard G arwin for his valuable criticism and encouragement. I cannot claim to be the first person to think of cake-sharing solutions to disarmament problems. The earliest mention that I can trace is by Singer [9]. The idea has been developed by Calogero [2] [3] [4].. The real credit must go to some long-forgotten child whose intelligence allowed him (or more probably her) to overcome a dangerous aspect of instinctive behaviour.

REFERENCES

[1] Apfel, R., Personal communication.

[2] Calogero, F., "Some remarks and a proposal concerning the limitation of strategic armaments", Procs. 22nd Pugwash Conference, pp 305-317, Oxford, September, 1972.

[3] Calogero, F., "A scenario for effective SALT negotiations", Science and Public Affairs, pp 17-22, June 1973.

[4] Calogero, F., "A novel approach to arms control negotiations?", Procs. 27th Pugwash Conference, pp 1-10, Munich, August 1977.

[5]   Dublins, L.E., and Spanier, E.H., "How to cut a cake
      fairly", American Mathematical Monthly, Vol. 68,
      pp 1-17, 1961.

[6]   Garwin, R., Personal communication.

[7]   Gorovitz, S., "When both bidders lose at a stupid game',
      International Herald Tribune, 16th August 1983.

[8]   Rebman, K., "How to get at least a fair share of the
      cake", Mathematical Plums, Chapter 2, pp 22-37,
      Honsberger, R. edt., Mathematical Association of America,
      New York, 1979.

[9]   Singer, E., "A bargaining model for disarmament
      negotiations", J. Conflict Resolution, Vol 7, pp 21-25,
      1963.

[10] Szilard, L., "The Mined Cities", Bulletin of Atomic
      Scientists, pp. 407-412, December 1961.

3. MODELLING BATTLE: LANCHESTER THEORY AND RELATED APPROACHES

Jympton: "Intelligence ... has established that the people attacking us are ... are ... are the enemy."

Bloodnok: "So that's their fiendish game, is it?"

Seagoon: "Gentlemen, do the enemy realise that you have this information?"

Bloodnok: "Oh no, we've got 'em fooled: they think <u>we're</u> the enemy".

Seagoon: "What a perfect disguise!"

(<u>More Goon Show Scripts</u>, Spike Milligan, Woburn Press 1973)

SOME PROBLEMS OF MODELLING BATTLE

D.R. Andrews and G.J. Laing
(Defence Operational Analysis Establishment, West Byfleet)

"Page after page of professional economic journals
are filled with mathematical formulas leading the
reader from sets of more or less plausible, but
entirely arbitrary assumptions to precisely stated
but irrelevant conclusions".

Letter from Prof. Leontief to American Magazine 'Science'.

1. INTRODUCTION

War is the ultimate form of conflict and the aim of this
paper is to discuss some of the problems in applying mathematics
in this area, based on the authors' experience over some years
at the Defence Operational Establishment at West Byfleet.

Perhaps a few words on operational research and on DOAE
would be helpful before going any further. Operational research
is a relatively new discipline. The term was probably first
used in connection with the work in 1937 of a small group of
scientists working on how the techniques of radar location of
aircraft might best be applied to the air defence of the UK.
The techniques of OR were extended and used to good effect
during WW II, particularly in the Battle of the Atlantic. It
was at about this time(1941) that the Army Operational Research
Group was formed at West Byfleet in an old country house taken
over from the Charrington family. In 1965 this became a tri-
service establishment charged with studies to fulfil the
all-embracing remit set out in a Defence White Paper soon after:

"Operational Analysis studies are essentially
concerned with objective choice between alternative
options. They have become increasingly applied to
questions of defence stategy and to the major
policy decisions which face NATO. They continue to
help in solving problems of resource allocation
within the Defence Budget, particularly in the
equipment field".

But perhaps a less pompous description of operational analysis
might be:

"The elimination of a little bit of sentiment and the
substitution of a little bit of arithmetic".

To fulfil this remit techniques are needed that can quantify,
in terms of some military measures of effectiveness, the
effectiveness and cost effectiveness of systems as well as the
mix and numbers of these systems required to provide a credible
deterrent against military threats to this country. To do this
DOAE has made use of a number of techniques which require
mathematics to some degree or other.

There are of course well-proven mathematical techniques
which can help in providing an efficient use of transportation
resources for reinforcement or resupply, of logistics, or of
maintenance and repair facilities. But the difficult area is
in trying to represent mathematically the processes which occur
in combat itself. This is the area that we will concentrate on
in this talk, focussing particularly on Army operations.

2.   OPERATIONAL CONTEXT

The main backcloth for British Army operations is that of
defending alongside our NATO allies, the Central Region of
Europe (Figure 1).

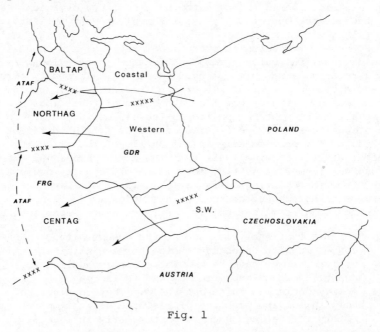

Fig. 1

It might be anticipated that, in any attack by the Warsaw Pact
forces, they would concentrate their forces in certain sectors
where the going is good for armour in order to penetrate,
envelop and defeat NATO forces. In trying to represent such a

scenario analytically, a major problem is the number of
equipments involved both in quantity and type. Thus for a
conventional conflict in the Central Region (1) the numbers of
forces and equipment could be as large as shown in Figure 2.
To make any progress it is clearly necessary to simplify and
aggregate.

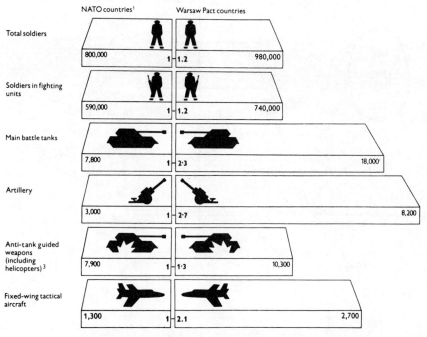

¹ Including French Forces in the Federal Republic of Germany but excluding the
  Berlin garrison, which is not declared to NATO
² Includes some Warsaw Pact tanks in training units and storage which would be
  available for operational use
³ Only weapons which are, or have the capability of being, vehicle or helicopter
  mounted are included

Fig. 2.

   Experience from previous wars suggests that armoured battles
are not continuous in time but occur as a series of large scale
actions followed by regrouping and redeployment. This is not
to say that nothing happens in the intervening periods but
certainly the largest losses occur in these intensive actions
and the outcome of the campaign is determined by the balance at
the end of a series of such actions. For this reason much of
DOAE's analysis has concentrated on force groupings of about
battalion level for Blue and one or two regiment size for Red.
Thus a stylised central region conflict might look rather as
shown in Figure 3.

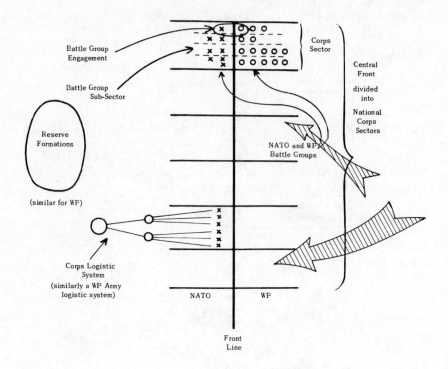

Fig. 3.

Given that the outcome of battalion level engagements can be
evaluated, the Central Region battle becomes one that can be
handled by an interactive simulation or war game using
aggregated results from the battalion level analysis as a
building block (Figure 4).

In view of its fundamental importance, considerable
attention has been paid at DOAE and elsewhere to the analysis
of battalion level engagements where one is looking at battles
which typically comprise some 200 armoured vehicles supported
by infantry, ATGW, minefields, artillery and air - still a
very complex battle.

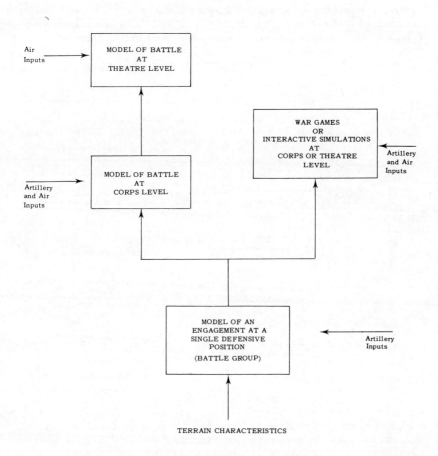

Fig. 4. A Hierarchy of Land Battle Models

3.   THE TECHNIQUES OF ANALYSIS

There are a number of ways of analysing such a battle all of which have advantages and disadvantages as indicated in Figure 5.

During a war, if it lasts long enough, there is usually an opportunity to acquire real operational data on which analysis can be based, and which can in turn be used to influence the future progress of the war. But such data are usually of a statistical nature, rather than of the detailed type which enables one to reconstruct the battle in sufficient detail to

validate analytical models. Indeed in researching back through
historical data it is difficult to be sure how many and what
type of forces on each side were involved and why events
occurred as they did.

CHOICE OF TECHNIQUES

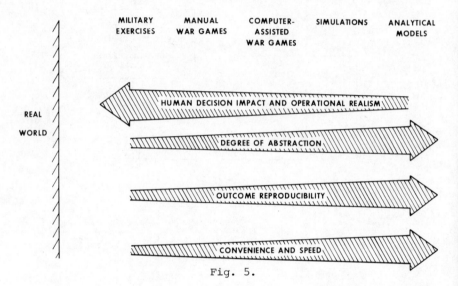

Fig. 5.

Data collection is never at the forefront of people's mind
when in battle, and our recent experience in the Falklands was
no exception. Moreover, of course, in mathematical terms, any
one battle only represents one possible outcome from a Monte
Carlo process. So we would really like the same battle
repeated perhaps 100 times!

But there is some good information we can obtain from
historical data particularly on suppression effects and on
defeat levels.

The next best thing is a fully instrumented tactical
interactive field trial. DOAE has the capability to mount
such trials and can obtain a lot of necessary basic data from
such trials to provide a credible input data to analytical
models and games. Such field trials are inevitably very
costly.

War gaming is an important tool when we wish to explore new
tactics or the use of new types of weapon. It is an essential
learning tool in those many situations where there is a need

to understand what to model and how to do it. But games are slow and can be dominated by player performance.

Simulations are really little more than war games with the decision making process automated with a series of 'if , then' rules. However they become complex if every piece of equipment and all the decision rules have to be represented. Nevertheless there has been a tendency for military analysis to develop along these lines partly because the developing power of digital computers has enabled large simulations to be handled. As a result simulations with thousands of variables have become commonplace, leading to problems both in feeding such simulations with data and in understanding and verifying the interactions taking place within the simulation. Moreover the simulations are often stochastic leading to a need for large numbers of replications to produce an average answer.

This 'crisis of complexity' problem is not new. It is discussed by Taylor (4) and was the theme for a US Army Operations Research Symposium in 1976 (2).

One way of overcoming this complexity problem is to use mathematical equations to represent the interactions and so produce a simpler model (or set of models) which can be run quickly so that a considerable number of variations can be made in equipment capabilities and mixes and in some of the basic assumptions that needed to be made about various factors.

4. USING LANCHESTER'S EQUATIONS

For such a 'fast' model use is often made of the equations that Lanchester first formulated in 1916. In their simplest form they can be written as:

$$\frac{dr}{dt} = - \beta b$$

$$\frac{db}{dt} = - \rho r$$

where

r is the number of RED weapons surviving at time t

b is the number of BLUE weapons surviving at time t

β is the rate at which a BLUE weapon can kill RED weapons

ρ is the rate at which a RED weapon can kill BLUE weapons

Use has been made of extended versions of these equations in the Battlegroup Model that has been in operation at DOAE since 1967. In their general extended form, and with coefficients which change as the battle closes, the equations can only be solved by step by step integration. However Tom Weale has been able to produce mathematical solutions for a 3 by 3 battle situation (3 categories of weapon on each side).

Papers specifically on Lanchester modelling have been provided for this conference by Peter Haysman and Trevor Lord (see below, this volume).

In DOAE's experience, mathematical models, such as those of the Lanchester type, can be elegant and simple but the real problem becomes the determination of the coefficients. It is all too easy for the mathematician to put some arbitrary coefficients into his equations, make some broad assumptions about the course of the battle and then derive some precisely stated but irrelevant conclusions. The problem with battles is that they are highly dependent on a large number of very uncertain factors some of which can be represented in analytical models and some cannot. Thus there is need for considerable sensitivity analysis looking at how outcomes will change with plausible variations in the input data and assumptions. Indeed it would be fair to say that it is generally the assumptions and data that drive the answers not the mathematics of the model.

The type of uncertainties involved can be illustrated by four examples:

(i) it will be obvious that the attrition coefficients in the Lanchester differential equation are an aggregate of a number of factors which vary through the duration of the battle. Thus if a target is moving towards the firer over real terrain it will be visible for discrete time periods and the length of these periods will dictate whether the firer has time to get 0, 1, 2 shots off. Hence rate of fire is dependent on both the terrain and the characteristics of the weapon. Whilst it is possible to derive an average value for a particular one-on-one situation, extending this to a range of terrains and to a group of firers against a group of targets clearly presents difficulty.

(ii) DOAE trials show that there are many occasions when a system already dead is fired at again. Many military officers would argue that with good training no overkill occurs, yet experience in tactical trials even with experienced troops shows otherwise. Figure 6 shows DOAE trial results (Rowland (3)) and shows that the overkill factor is about 2. Indeed as the right hand graph

shows this can be explained on the basis that any target's
probability of being hit is independent of whether it has
already been hit. Such a factor of 2 can make a big difference
to battle outcome.

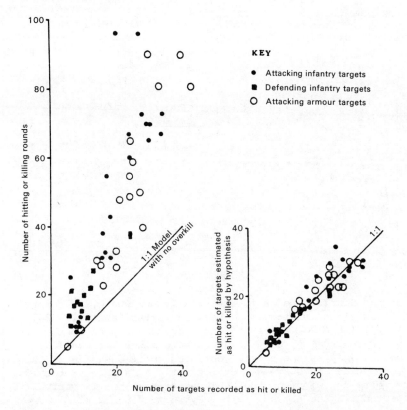

Fig. 6. Comparison of interactive trials data with conventional
        modelling assumption and simple hypothesis for overkill.

(iii) because weapons are expensive the numbers that can be
fired in trials are very limited. Moreover live firings have
to be against carefully presented targets for range safety
considerations. Thus there are considerable uncertainties in
extrapolating these data to determine operational performance
and reliability, particularly with the knowledge that, in war,
confusion, battle stress and fear are certain to degrade
operator performance. The problem is even more difficult when
looking ahead to the future, where the weapon systems on which
decisions need to be taken may be little more than 'paper
weapons' and our picture of war in the year 2000 uncertain.

(iv)   what does one assume about defeat levels? Figure 7 shows
a diagram produced by Tom Weale (5). It illustrates the large
differences in the casualties sustained when the battle ends.

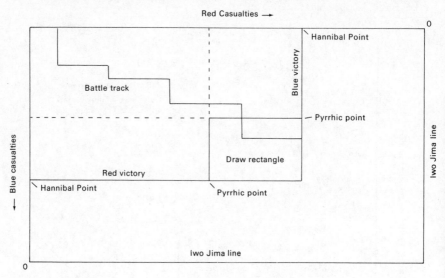

Fig. 7.

Thus the Hannibal point is essentially one where, by surprise,
an overwhelming victory was obtained (after Hannibal c 247-183
BC). At Iwo Jima (Feb 1945), the Japanese fought to the last
man.   More usually, casualties to both sides increase
progressively as the battle continues until one side or the
other decides to surrender or retreat. A Pyrrhic victory is
one where the winning side achieves victory but is virtually
wiped out in the process (after Pyrrhus c318-272 BC). Surprise
apart, morale and motivation can play a big part in determining
the casualties at which an attack will be pressed home or the
defender retreat or surrender. Figure 8 shows some historical
analysis by David Rowland of battles where infantry are in
the attack. The casualty levels at which there is a 50:50
probability that the attack will be pressed home vary
considerably. In the Boer War British attacks went to ground
at a median of 10% casualties, probably because of an
unwillingness to die to such a remote conflict. On the
Somme in WW I preliminary data suggest that 32 of 81
attacking battalions persisted in attacks with over 60%
casualties, but attacks later in the war were not pressed in
the same way. WW II data are intermediate.

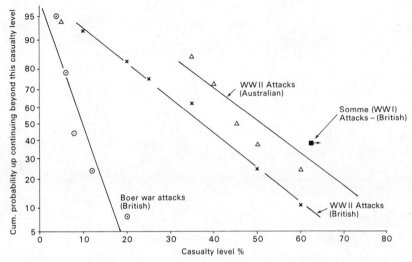

Fig. 8. Defeat levels in infantry attacks

5. CONCLUSIONS

In summary mathematics has a useful role to play in the representation of military conflict, but care has to be taken not to become so dazzled by the elegance of the mathematics that one loses sight of the importance of input data and assumptions and of the problems that are left to others to massage the input data into the form needed by the mathematical equations. We are, after all, looking for robust solutions over a wide range of uncertainties rather than carefully optimised solutions applicable to only one set of data and assumptions.

REFERENCES

1. Statement on Defence Estimates Cmnd 9227. 1984.

2. Harrison, D.C., 1976, 'Keynote Address' pp 1-20 Proceedings of the Fifteenth Annual US Army Operations Research Symposium, Fort Lee, Virginia.

3. Rowland, D., 1984, Paper presented to the International Symposium on Advances in Combat Modelling. RMCS Shrivenham.

4. Taylor, J.G., 1980, Force-on-Force Attrition Modelling Military Applications Section/Operations Research Society of America (MAS/ORSA)

5. Weale, T.G., Stochastic Lanchester Theory using Numerical Methods. Paper presented at the 1979 Annual Conference of the Operational Research Society, Stirling.

THE ROLE OF OPERATIONAL ANALYSIS

T. Price
*(Formerly Director, Defence Operational Analysis
Establishment, Byfleet)*

INTRODUCTION

To someone like myself who was engaged in defence analysis
twenty years ago, a major source of interest in this conference
is how far the subject has advanced, and whether all the modern
techniques and computing facilities have made mathematical
analysis more useful as a tool at the level of central defence
policy-making. The dominant and rather surprising impression
which has emerged from this day of erudition, much of it rather
abstract, is how recognisable the problems still are, how
similar the conclusions are to those which we reached
empirically twenty years ago, and how little has fundamentally
changed. The key points of the 1960s remain the key points of
today, despite the vast alterations which have occurred
elsewhere in the defence field.

Within these college walls it is appropriate to start with
a favourite warning of the Master, Sir Hermann Bondi: watch the
assumptions. People, he says, can usually be relied on to do
their arithmetic correctly. But if the result is surprising,
or even absurd, one would be well advised to look at their
assumptions BEFORE Equation One.

Today's defence budgets are limited in cash terms. If the
Navy gets more money, then the Army or the RAF necessarily get
less: at this point, at least, we are involved in a zero-sum
game. So we can rely on the Services marking each other and
watching the assumptions. As an earlier speaker remarked, when
playing the budgetary game the trick is to find a set of
assumptions that will give the "right" answer. What is "right"
naturally depends on one's professional perspective. But at
least we need not doubt that there will be discussion within
Whitehall about assumptions. Taking a range of assumptions
offers at least some protection against Richardson's difficulty

- that translating verbal statements into mathematical terms
can too easily lead to a spurious air of definiteness.

SOME EXPERIENCES, OBSERVATIONS AND LESSONS

Twenty years ago I remember the debate about the Air
Force's island-hopping strategy: a hoped-for ability to take
reinforcements to almost anywhere in the world at three days'
notice, by the massive use of air-power. At the time we were
still engaged in the Far East, and one of the possible options
was to use a small island in the middle of the Indian Ocean, as
a staging post. If that happened, then within three days of an
alert, according to the simulations, the island would reach a
peak of activity comparable to that at Heathrow on a summer's
afternoon. By the merest chance, one of those taking part in
the debate had spent a night during World War II on the island,
and knew it to be a small, narrow, and slightly curved coral
atoll. The simple question, whether there would be enough room
for hard-standings, brought one phase of the discussion
virtually to a close.

The larger the simulation, the more assumptions go into its
making, and the more one loses the intuition which can come
from smaller, more sharply focussed, pieces of modelling. It
becomes increasingly difficult to know whether what one is
doing is sensible. Blackett, the Noble prize-winner whose
operational research contributions were so important to the
Navy during World War II, advised me when I first went to
Byfleet to avoid large simulations for just that reason.

We did not always take his advice. Once we were doing
some work on the speed with which RAF Transport Command could
be 'generated' - to use the jargon word - for action. A large
computer simulation was accordingly begun, correct right down
to the level of individual crews. It grew and grew, until cost
forecasts for programming time reached a quarter of a million
pounds - and they were 1965 pounds. But just as we were
beginning to despair of ever reaching a conclusion we recruited
Gerry Lorimer, who by a fortunate chance had been in charge of
operational research at Transport Command itself. He brought
with him the intuition we lacked and so badly needed. Within
three weeks the problem had been solved by very simple and
direct simulation at virtually no cost, because he knew where
the key elements lay. There is a lesson in this for those
building up analysis teams: mixed experience and mixed disciplines
are essential if a great deal of wasted time is to be avoided.

Simulations of all kinds are necessary in peacetime to
provide the synthetic experience of military conflict which
could otherwise be provided only through very expensive and

often not very realistic field trials. But, as Alan Goode, one
of the Byfleet superintendents, pointed out, simulation results
may depend very sensitively on the numbers which are used to
describe the capabilities of the various units and weapons
systems. Those numbers may be no better than "guestimates";
but they may be the only thing available, and in that case we
face the danger that a whole pyramid of defence policy
argumentation could be built on them. After a time they might
even acquire the status of established facts, almost
unquestioned in further debate. If they were wrong or
misleading this might not be discoverable from conventional
military exercises, which by their nature tend to be broad in
their coverage, and not concerned with the exact values of
individual parameters. Being often pre-programmed, such
exercises also fail to provide that element of chance which
plays such a large role in actual combat. Chance can of course
be explored via repeated Monte Carlo simulations. Goode argued
that it would be only sensible to try to support such
simulations with specially designed experiments, which would
study the 'essence' of the matter, so that we could then have
some hope of deducing the values of the critical parameters.
Such experiments would be wholly different from ordinary
military exercises, more like the clean experiments of physics.

    At that time we were still in the early stages of
exploring the possibility that helicopters armed with guided
weapons might be able to tip the balance of tank warfare in our
favour - by flying tactically, rising up to spot an enemy tank,
loosing off a guided weapon, and sinking down again behind
trees or other cover. What we did not know, and could hardly
infer from a process of pure thought, was how easy or how
difficult it would be to acquire the enemy tanks as targets;
how easily NATO helicopters could in turn be acquired by the
enemy - when with modern weapons, with their high hit
probability, they might well be virtually dead; or how
successfully they could avoid similar treatment at the hands of
friendly forces. So a highly stylised experiment was mounted
to determine those data, using two different and fairly easily
distinguished types of helicopter, to represent NATO and Warsaw
Pact aircraft. The results confirmed the value of such an
approach: they contained some surprises, not least a
significant proportion of wrong identifications. Such results
could not have been obtained by any other means.

    Besides assumptions and parameters, there are also the
criteria which influence the taking of decisions. The case of
tank versus tank combat involves a balance between gun calibre
and range. The bigger the gun the further off the enemy can be
hit; but fewer potential kills can then be stowed on board.
Since with modern ammunition kill probabilities are - or were

before the days of Chobham armour - uncomfortably high, the
criterion for optimisation is not immediately obvious. Should
you accept anything less than the maximum range, which might
enable you to out-reach the opposition before he could hit you,
rather like the battles between capital ships during World War
I? Or, in view of the difficulty of spotting a well-
camouflaged tank in broad daylight even at a distance of no
more than 1500 yards, well below gun range - would it be better
to concentrate on maximising the number of rounds and speed of
firing? This will be recognised as a particular case of the
quality versus quantity argument, which has never been, and
probably never will be, properly settled.

Only rarely is it possible to argue from pure logic.
Blackett's large convoy theory is a case in point. The number
of ships a World War II convoy could contain was proportional
to the area it occupied, while the length of the defensive
perimeter increased more slowly than the number of ships. So,
even though it means putting many valuable eggs in one basket,
it should be advantageous to rely on large, well-defended
convoys. The argument is not completely rigorous, because it
does not deal with the situation where, in spite of all one's
efforts, the perimeter defence is broken. The stakes being huge,
Blackett agonised over his recommendation, finally reaching a
decision by asking himself whether he would be happier if his
own son were in a large convoy. Answering his question in the
affirmative, he went ahead with the recommendation.

In the early 1960s we were still coming to grips with the
issues of nuclear war - particularly the opening phases, which
it was essential to deter. NATO had its dispositions, but
there was no framework of established defence philopsophy on
which to rely for this task of maintaining peace. We seemed to
be on our own. But there was one book which came to have a
seminal influence - Tom Schelling's 'Strategy of Conflict'.
He demonstrated, partly by simple game theory, partly in words,
that if you wish to deter a rational opponent you must place
yourself in a position where, if he makes a move, he will
understand that you may not be able any longer to control the
conflict and limit the damage suffered by either party, even
if you wished to do so. The outcome would be so unpredictable
that a rational opponent must be deterred. So Tom Schelling
argued. This valuable insight proved to be just what was
wanted. Schelling's argument pointed firmly in one direction -
that there must be no "fire-break" between a platoon incursion
and the strategic deterrent; otherwise the enemy might see, or
believe he could see, a possible route for escaping
destruction. Schelling's formulation provided a meeting point
for ideas, avoiding all the theology of 'trip wires' and the
rest of the jargon. Fortunately, NATO's dispositions at the

time seemed to be reasonably consistent with what was needed.

Extensive simulation and mathematical modelling were of
course used to explore what might happen if the enemy did not
behave logically, was not deterred, and advanced into Western
Europe. The results differed from simulation to simulation,
but one disturbing feature was always present. The conclusion
appeared to be that, given the weapons available 20 years ago,
it might have been difficult to halt the enemy except at the
cost of launching a large number of tactical nuclear weapons.
Well-equipped enemy forces marshalled in appropriate tactical
formations are much less vulnerable than the people and
buildings of the territory through which they are advancing.
In the words of Sir Solly Zuckerman, then Chief Scientific
Adviser at the Ministry of Defence, one might be facing the
possibility of 20 Hiroshimas between lunch and tea on a narrow
front. But the resulting holocaust would then come close to
sweeping away the political purposes for which the war was
being fought. It was considerations like these which led
Mountbatten to conclude that nuclear weapons were not for
fighting wars, only for deterrence. Zuckerman gave a speech at
Shapex in the very early 1960s saying as much (it was later
published in Foreign Affairs). It was the act of a brave man.
It aroused a furore, including questions in the House of
Commons. The Prime Minister, MacMillan, was splendid, and gave
the necessary support. Fortunately military analysts do not
always have to take such exposed positions; but there are times
when they have to nail their colours to the mast. There is
still a great deal of validity in the conclusions reached at
that time, and Zuckerman's speech has hardly been dated by the
passage of time.

There was another case in my time at Byfleet where
prolonged and detailed analysis led to the comforting
conclusion that the existing force mix, built up piecemeal over
the years, was not far from what might have been considered
optimal for the military and political purposes of the time.
In the mid-1960s we had not yet extricated ourselves from the
Middle East, and were still interested in exerting a military
influence in the area of the Arabian peninsula.
This particular issue was whether a commando ship
needed to be replaced. The question first had to be re-cast in
researchable terms. The key was the degree of warning time
which one might expect to receive before an incident. Those
were still war-like days: we had experienced more than 60
incidents since World War II, and there was a substantial body
of information on which to draw. Given sufficient warning time
the Navy could be pre-positioned over the horizon. Shorter
warning times would favour the rapid response of the Air Force.

Army garrisons could be there on the spot, but would need
reinforcement. Each Service, and each type of unit, had
certain capabilities which could be expressed mathematically in
the form of simultaneous equations.

Having as a result of a preliminary analysis concluded that
in the circumstances of those days the probabilities were
strongly in favour of receiving adequate warning, Nigel Beard's
team constructed a linear programming model of over 700 lines.
The solutions had to be integral, half a ship or aircraft not
being much use. The problem, the largest of its kind up to
that time, was successfully solved, and the solution pointed to
a force mix similar to that which was already planned. The
fact that analysis had on this occasion given the same
conclusions as those reached by other means did not mean that
the time and effort had been wasted. The robustness of the
solution had been demonstrated over a wide range of scenarios.
The case for the commando carrier had been clarified. And a
valuable new approach had been created, which was later able to
identify previously unsuspected bottlenecks in NATO's logistic
plans.

Operational analysis sometimes has to work on assumptions
which might be outdated in time by clever engineering. On
paper the aircraft which eventually became the Harrier was a
non-starter, with virtually no pay-load. But the Ministry of
Aviation and its successors refused to be discouraged, with the
result we saw in the Falklands. There were of course failures,
like the TSR2. By overstating the Operational Requirement in
terms of the length of the supersonic low-level dash capability
- which was at the very edge of what was technically possible -
the size and unit cost were forced up to a point at which the
aircraft, although a promising flying machine, became a
political football and was cancelled. A greater degree of
realism in formulating the operational requirement might have
led to a different outcome, and avoided the subsequent purchase
of Fllls.

This example is a reminder that very hard-headed economics
must not be overlooked as an essential input to certain kinds
of military operational analysis. There have been far too many
overspecified and therefore impossibly expensive defence
projects, which have inevitably led to cancellations. As
technology advances it will continue to raise the cost of 'the
best', and make choices ever more difficult. The quality
versus quantity argument can only become more intense.

Mention of operational requirements leads me to recall
that the change of name from operational research to operational

analysis in the early mid-1960s was not, as I have heard argued by a distinguished academic in all seriousness, because of any change in subject content; but simply for administrative convenience at the time of the 1964 Defence White Paper, to avoid confusion between the two kinds of OR, when abbreviated.

CONCLUSIONS

Where do these reminiscences of twenty years ago leave us? With a feeling that, for all the changes in weaponry, and the vastly increased computing power available to the analysts, the position of operational analysis today as a servant of defence policy-making has not greatly altered. Such analysis may be directly useful only in rather rare circumstances, it being easier to explain the past than to predict the future. But it is indirectly useful in a variety of ways. It may offer comfort and support for decision-making carried out by other means. Or it may raise queries, which may be even more valuable. It generates fresh insights. Perhaps most important of all, it forces the analyst to cast the problem in researchable terms: which is itself a most powerful step towards identifying the essence of a problem. And it focusses attention on assumptions, parameters, and criteria - things which might be taken for granted were it not for the explicit nature of the analysis. If there is a trap it is to imagine, wrongly, that casting knowledge and facts into mathematical and numerical form thereby encompasses the whole truth. As Denis Healy once remarked when Secretary of State for Defence, we should remember that there are also facts which can only be expressed in words.

The need remains today, as always, for the analyst to have a powerful patron. Should he find that his analysis confounds accepted wisdom he will need all of that patron's support, and confidence in his own correctness, if he is to carry his point against the pressures that can build up in a major Department, where the decision he is challenging may inter-mesh with countless other considerations. But in the end the truth stands a very good chance of prevailing. The Ministry of Defence is one of the most objective of all government Departments, and firmly wedded to the principle of applying intellectual rigour whenever possible to policy-making. The motivation is immediate and clear-cut: the men of the Armed Services might themselves be the first victims if the arguments are not right. It is splendid territory for operational analysts, now as in my own day.

REFERENCE

Schelling, T.C. (1960): "The Strategy of Conflict" Harvard
University Press (2nd edition 1980).

ACCURACY AND LANCHESTER'S LAW:
A CASE FOR DISPERSED DEFENCE?

Robert Neild
(Trinity College, Cambridge)

As a non-mathematician, it is with the greatest
hesitation that I offer a paper to this conference. My excuse
for doing so is that I have come across a problem relating
to Lanchester's Square Law. It concerns the validity in
today's conditions - and tomorrow's - of the assumptions
underlying that law. I think the point merits attention for
two reasons. First Lanchester's Square Law seems to be quite
widely used. Secondly, the change in assumptions with which
I am concerned may be of some help in the development of
sub-nuclear strategy, a subject to which attention is now
turning as the limitations of extended nuclear deterrence are
perceived. I must emphasise that the discussion that follows
is concerned with non-nuclear weapons.

1. LANCHESTER'S LAW

Lanchester put forward two related propositions, as
indicated by the title he used. It was "The Principle of
Concentrations. The N-Square Law" ([6], Chapter V)

He based the principle of concentration on the conditions
of what he, 70 years ago, called "modern warfare" contrasting
it with "ancient warfare". He said,
"In olden times, when weapon directly answered weapon,
the act of defence was positive and direct, the blow of
sword or battleaxe was parried by sword and shield, under
modern conditions gun answers gun, the defence from
rifle-fire is rifle-fire, and the defence from artillery,
artillery. But the defence of modern arms is indirect:

I am indebted to Julian Hunt for comments and to Bob Rowthorn
for Appendix 1 and for comments.

tersely, the enemy is prevented from killing you by your
killing him first, and the fighting is essentially
collective. As a consequence of this difference, the
importance of concentration in history has been no means
a constant quantity. Under the old conditions it was not
possible by any strategic plan or tactical manoeuvre to
bring other than approximately equal numbers of men into
the actual fighting line; one man would ordinarily find
himself opposed to one man. ... Under present day
conditions all this is changed. With medium long-range
weapons - fire-arms, in brief - the concentration of
superior numbers gives an immediate superiority in the
active combatant ranks, and the numerically inferior force
finds itself under a far heavier fire, man for man, than
it is able to return. The importance of this difference
is greater than might casually be supposed, and, since it
contains the kernel of the whole question, it will be
examined in detail" (op. cit. pp. 40-41).

Lanchester then shows how in ancient conditions of direct
man to man combat it made no difference whether you split your
forces or concentrated them ie. whether you enjoyed numerical
inferiority or superiority in each engagement. The outcome
depended on the relative total size of the two opposing armies -
and the "fighting value" of the men in the two armies.

To investigate his "modern conditions" he assumes that two
forces, "Red" and "Blue", face each other on equal terms,
meaning that the soldiers have the same "fighting value" and
fight in symmetrical conditions as regards cover and other
variables, so that each man will in a given time score, on an
average, a certain number of hits that are effective. Using b
to represent the numerical strength of the Blue force and r for
the Red force he writes

$$\frac{db}{dt} = - \rho r$$

and

$$\frac{dr}{dt} = - \beta b$$

where ρ and β are constants for the fighting values of the
soldiers on the two sides. Lanchester uses graphs to
illustrate the advantages of superiority (I shall do so with a
numerical example below), and then arrives at his famous square
law as follows. He introduces a condition of equality between
the two sides, defined as the condition "when in combat their
losses result in no change in their numerical proportion", ie.

when the strength of both sides declines at the same
proportional rate. In other words.

$$\frac{db}{b.dt} = \frac{dr}{r.dt}$$

Suppose that for this to hold, the fighting values of the men
in the Blue and Red forces (β and ρ) take some values M and N.
The rate of reduction of the two forces is then given by,

$$\frac{db}{dt} = -Nr \; ; \; \frac{dr}{dt} = -Mb$$

Substituting, he gets

$$-\frac{Nr}{b} = -\frac{Mb}{r}$$

$$\text{or } Nr^2 = Mb^2$$

In other words, the fighting strengths of the two forces are
equal when the squares of the numerical strength multiplied by
the fighting value of the individual units are equal. (In the
limit, the two sides would fight to mutual annihilation).
More generally, the law says: the fighting strength of a force
may be broadly defined as proportional to the square of its
numerical strength multiplied by the fighting value of the
individual units. (p.48).

It is impossible not to admire the beauty of this law. It
provided a brilliant rationale for concentrating your forces -
and dividing those of the enemy if possible - so as to achieve
superiority of fire-power, something which military men knew
from experience to be of the greatest importance. And it
provided a rule of thumb for predicting the outcome of battles -
if the relevant assumptions were reasonably well fulfilled.

2. CRITIQUE OF THE ASSUMPTIONS

In order to see precisely what lies behind the law and
examine the assumptions critically, I find it helpful to use a
numerical example.

Suppose two forces of soldiers, again Blue and Red, start
shooting at each other in symmetrical conditions (eg. with the
same rifles and standards of aiming) so that the hit
probability (a term, I use in preference to Lanchester's
"fighting value" for reasons explained in Appendix 2) is the
same on each side. The only thing that is asymmetrical is the
number of men. Suppose there are twice as many Blues as Reds.
At the first round of fire, (supposing they start firing
simultaneously) the Reds can aim at only half the Blues. But

on the other side two Blues can aim at each Red. The result is
that the Reds, who were fewer to start with, lose nearly twice
as many men as the Blues. Consequently, at the second round of
shooting, the ratio of Blues to Reds will be even more
favourable to the Blues than at the first round. There will be
only enough Reds to aim at less than half the Blues; and there
will be more than two Blues to aim at each Red. The difference
in casualty rates will, therefore, be greater than at the first
round and the ratio of Blues to Reds will show a further and
bigger increase in favour of the Blues. In this way the
differential advantage to the Blues will accelerate
dramatically as shooting continues.

To construct the mathematical example shown in Table 1 I
make these assumptions:

a) 2,000 Blues fight 1,000 Reds. (Large numbers are used
    so as to avoid fractions of a man.)

b) The kill probability per shot is 0.2 on both sides.

c) At each round all the Blues and Reds shoot
    simultaneously and the bullets cross in mid-air. (In
    order to isolate and examine the consequences of
    numerical superiority and nothing else, we rule out the
    possibility that one side or other may gain by shooting
    first; and we rule out the possibility that the
    numerically superior side may stagger their fire within
    each round).

d) Each man on each side at each round is given a target,
    selected on the basis that fire is allocated evenly to
    targets so as to maximise total kills. (If it is
    assumed that each man chooses his own target and target
    selection is random, the rate of kills will be less than
    in this example. The extent of the difference can be
    seen by comparing the numbers generated here (Table 1)
    with those generated in Appendix 1 by Mr. Rowthorn, who
    assumes that aiming is random but that everything else
    is the same as in this numerical example).

The figures can be explained as follows:

a) In Round 1, 1,000 Reds aim at 1,000 Blues and kill
    1,000 x 0.2 = 200 Blues. The number of Blues surviving
    at the beginning of Round 2 is 2,000 - 200 = 1,800.

b) In Round 1 on the other side, 2,000 Blues aim at 1,000
    Reds, spreading their fire evenly so that they aim two
    shots at each Red. Consider their shots as two waves,
    even though they are, by assumption, simultaneous.
    The first wave of 1,000 shots will hit and kill
    1,000 x 0.2 = 200 Reds. The second wave will

hit 1,000 x 0.2 = 200 Reds. But because 200 of 1,000
Reds will have been killed by the first wave, the number
killed by the second wave will be only 800 x 0.2 = 160.
So the Red casualties in Round 1 are 200 + 160 = 360.
The number of Reds surviving at the beginning of Round
2 is 1,000 - 360 = 640. The ratio of Blues to Reds at
the beginning of the second round is 1,800 to 640 or
2.8 to 1.

c) In Rounds 2, 3, 4 and 5 the process continues with the
    ratio of Blues to Reds taking off dramatically. Every
    shift in that ratio means fewer Blues will be shot at
    and that more shots will be fired at each surviving Red.
    These are two sides of the same coin - the coin which
    generates the accelerating change in the ratio.

Table 1

Round	Blues		Reds		Ratio of Blues to Reds
	Number at beginning of round	Casualties during round	Number at beginning of round	Casualties during round	Ratio at beginning of round
1	2,000	200	1,000	360	2:1
2	1,800	128	640	297	2.8:1
3	1,672	69	343	227	4.9:1
4	1,603	23	116	111	13.8:1
5	1,580		5		316.0:1

This law can be applied equally well to an artillery duel,
a tank battle or other forms of exchange; and the same
conclusion follows, namely that it pays to concentrate your
forces for the sake of numerical superiority.

However the law depends on two implicit assumptions. The
first and more obvious is that the advantages of being on the
defensive, with dug-in positions, knowledge of the ground and
so on, are assumed away, as are the advantages the attacker may
enjoy if he achieves surprise. The second implicit assumption,
whose significance is less immediately obvious but much more
interesting, is that the hit probability of the weapons is fairly
low, so that repetitive shooting is needed for an assured hit.

Suppose the hit probability was one, meaning that every shot
was sure to destroy its target. There would be no point in
numerical superiority. The Blues, with their first shots,
would kill one thousand Reds. There would be no point in
having more Blues to kill each Red more than once. One
thousand Reds would simultaneously kill one thousand Blues
with their first shots, if the bullets of the two sides
crossed in mid-air. Of course that would not happen. The
fact that it would not happen is a key point; it brings us to
a conclusion: as kill probabilities rise, the value of
numerical superiority declines and the value of shooting first
increases.

That this conclusion may be relevant to present and future
sub-nuclear warfare is pretty clear. The trend of technology
is leading to increased accuracies, and seems likely to
continue to do so, as improved sensors and guidance systems
are applied to projectiles of many kinds. The other variable
that matters is the lethality of sub-nuclear munitions if they
hit their targets. I imagine this is probably rising but not
as much as accuracy. Provided it is not falling so much as to
offset the rise in accuracy, the trend in kill probabilities
will be upwards.

I know that people argue about the efficiency of today's
precision guided munitions but what matters is whether they
would dispute that there is a trend reaching into the future,
possibly a strong one, to greater accuracy and higher kill
probabilities per shot. This seems most pronounced as regards
firing against military vehicles - armoured vehicles, soft
vehicles, aircraft and ships - since they present a good
"image" to sensors. (See for example [3], in which the term
"manned platforms" is used, rather than "vehicles"). Against
persons, I imagine that the improvement in accuracy may be
less important. On the other hand, the improvement in the
lethality of sub-nuclear area weapons has increased, so that
if a concentration of soldiers outside vehicles can be located,
the chances of knocking them out must have increased.

But let us pursue the details of weaponry no further and
simply take as assumptions, for purposes of theoretical enquiry,
the following points:

a) There is a sharp continuing trend to higher accuracy
    and lethality per shot against vehicles - and against
    persons if they can be located.

b) Kill probabilities per shot will approach unity.

3.   THE LAW OF DISPERSION

These assumptions have the following implications. Who
sees first, whether by eye or man-made sensor, kills.
Therefore concealment, from the eye and sensors, pays.
Since dispersion helps concealment, dispersion pays. Thus
we are led in a few steps to the opposite conclusion from
Lanchester. In his low accuracy world, concentration pays.
In our high accuracy world, dispersion pays.

In mathematical terms, we could write that the chances of
survival were a function of two probabilities, the probability
of being detected (P_d) and the probability of being killed once
you had been detected (P_k).

Thus, chance of survival = $(1 - P_d)(1 - P_k)$

Lanchester implicitly assumes that there is not problem of
detection: $P_d = 1$ and the chance of survival in an encounter
depends on $(1 - P_k)$. I am explicitly assuming that $P_k = 1$ and
the chance of survival in an encounter depends on $(1-P_d)$.
To minimise P_d, I go for dispersion.

This argument needs refinement and qualification. We need
to address the question, what degree of dispersion is required
and how does this apply to the defender and the attacker?
To answer that properly, we would need to bring in many
considerations other than accuracy alone; we would need
modelling and manoeuvres. But let us see how far we can get
with our bare hands - or brains. The first answer to the
question, "What is the optimum degree of dispersion"?, would
be "The degree that maximises concealment".

A second reason for dispersion is to reduce the
vulnerability of men and weapons to bombardment with area
weapons, nuclear and sub-nuclear, delivered by artillery,
manned aircraft or missiles. We can call this dispersion for
purposes of invulnerability. It must be brought into our
calculus alongside concealment.

At this point we must distinguish attacker from defender,
remembering that the defender may go over to the attack to
recover territory, and the attacker go over to the defence of
the territory he has taken. The attacker has to expose himself
in order to move forwards and occupy territory. It is hard
for him to conceal himself - though he may attempt to do so
with smoke, camouflage and other devices. On the other hand,

the defender, if he is appropriately positioned, need not
reveal himself until he fires - and even then remotely-controlled
or self-triggered devices (eg. mines) may be used, so that the
soldier is not revealed. Clearly, improved accuracy, which
rewards concealment, rebounds to the relative advantage of the
defender. He has the greater scope for concealment.

Thus the inherent advantages of defence, which have long
been recognised, are enhanced by improved accuracy. In other
words, the value of P_d has normally been higher if you are
attacking than if you are defending, a point not recognised in
Lanchester's law; and with increasing accuracy that difference
is going up. What are the implications for the deployment of
forces by the defender and the attacker respectively?

In the case of the defender, the maximisation of
invulnerability and concealment is likely to lead us to
uniform dispersion of defensive forces throughout the area to
be defended. This will not be so if the area is so sparsely
populated that optimum dispersion is achieved while leaving
some areas naked. (Australia is an extreme example). And
variations in the geographical possibilities of concealment
may make it efficient to deviate from uniform dispersion. So
may variations in the value of particular bits of territory
eg. headquarters or a capital city. But the achievement of
invulnerability and concealment drives one towards uniform
dispersion, subject to a limit when density tails away. It
drives one towards dispersed absorptive defence.

As regards the attacker, I find it harder to draw lessons.
Of course, if he can achieve surprise, so that the defence is
not ready, he gains. But that has always been true - and the
gain will be less against dispersed defence than against
linear defence. So let us abstract from surprise and first
consider attack with sub-nuclear forces, consisting of vehicles,
men and instruments of bombardment, against a defence
consisting of dispersed accurate weapons against vehicles and
persons.

If the defence is uniform over the area to be attacked
(eg. the whole of a small country) and each defending unit
can knock out a given number of attackers before being
over-run, the area the attacker can occupy will, in the
rock-bottom case where we abstract from all complications, be
the same regardless of how he attacks. If he attacks all along
the frontier in uniform strength, he will take a wide, thin
slice of territory before his forces are expended. If he
concentrates his forces on a narrow front he will take a narrow
slice reaching deep into the invaded territory, but the area

will be the same. We have the pure case of defence by attrition.
Which strategy will be best for the attacker? There is nothing
to tell you.* Rather the answer will depend on the political
objectives of the attacker. If his aim is to achieve political
submission by his opponent, he will probably go for the deep
slice, aiming to take his enemy's capital or to cut the country
in half. To guard against that eventuality, the defender can
deploy long-range bombardment systems - say smart short - to
medium-range missiles - with which to destroy and break-up
concentrations of enemy forces. These missiles too would need
to be dispersed, concealed and perhaps mobile, for the sake
of invulnerability.

In this way the logic of the argument leads us to a
two-tier dispersed defence consisting of short-range accurate
weapons supported by long-range accurate weapons to break up
concentrations of attacking forces. This is essentially
the model evolved by Dr. Afheldt and his collaborators in
Germany. [1]

It may well be objected that so long as the other side has
offensive forces the defender must have a counter-attack
capability. Part of his defence budget should go on forces
with an offensive capability, part on the defensive network.
But how far one should go in that direction is not at all
obvious. The more a given defence budget is devoted to
offensive forces, the greater the territory you are likely to
lose if the enemy attacks, since your defensive network will
be thinned - and your offensive forces cannot be relied upon
to meet and defeat the invading army. On the other hand, the
greater will be your chances of driving him out - or of
taking a slice of his territory as a bargaining chip. But on
top of those considerations there is another that is perhaps
more important, at least in the long run.

Suppose, as seems to be happening in significant degree,
technology offers a choice between forces which in their
weapons deployment, doctrine, training and logistics are
strong in defence and forces which, conversely, are strong in
offence. Suppose further that two countries, whose aims are
defensive, seek to achieve peace by the possession of armed
forces. The more they go for offensive forces, the more they
will generate instability, by provoking in the minds of their
opponents a fear of attack, possibly a temptation to attack
pre-emptively, and a feeling that they must arm competitively
ie. an arms race. The more they go for defensive forces, the
more they will generate stability, by removing the fear of

*There is one important qualification. The proposition that the
area taken will be the same if, ceteris paribus, the attack
comes on a wide or narrow front is true only if the fields of
fire of the concealed defending units do not overlap. If, as
likely, they overlap, the narrow deep attack will suffer more
casualties from fire from overlapping units on its flanks than
will the wide shallow attack. So the area taken will be less.

attack from the minds of their opponents, by removing the risk
of pre-emptive attack and removing the pressure to arm
competitively.   Indeed they may induce a reciprocated reduction
in arms: Blue goes defensive, Red can cut its forces, whatever
their mix of defensive and offensive capability, and be as safe
as before.   If Red at the same time goes defensive Blue can
reciprocate, and so on.   A virtuous spiral might be induced
by changes in strategy and force structure.   This, as Anders
Boserup has pointed out [2], is a more promising way out of an
arms race than any amount of negotiation by nations fearfully
pointing offensive arms at each other.   In more technical terms,
if forces were as strong in attack as in defence, as Lanchester
implicitly assumed them to be, any balance of forces would be
unstable: if one side got a lead, by design or accident, it
could rely on cumulatively overwhelming its opponent.   But as
there are advantages in defence, as there appear to be
increasingly, stability increases.   A question with which I am
concerned, in work I am doing jointly with Boserup, is how far
one can cultivate defensive superiority, using modern
technology, so as to increase stability.

   This has taken us rather far from Lanchester and the
reversal of his case for concentration.   My objective has
been to pursue the implications of improved accuracy for
defence by dispersion.   To return closer to Lanchester, the
main question I would like to put to my mathematical colleagues
is this:

   Can one produce any simple rule to say at what level of
   accuracy it pays to disperse rather than concentrate?

   My amateur view is that one cannot provide a simple answer.
It is possible, using the kind of simplifying assumptions I
have made, to calculate what proportion of a country an
attacker with a given number of units would occupy if faced by
a given number of dispersed defending units achieving a given
rate of first shots at given kill probabilities.   But you then
have to compare that with what happens if the defender
concentrates his defending units, and that is a very different
kettle of fish.   You can never know in what relative strength
the two sides will meet since, with concentrated forces, the
game which both sides will be playing is to out-manoeuvre
the other side.   And you have the problem of the vulnerability
of concentrated forces to area bombardment.   I am therefore
led to the view that you have to bring in wider considerations
and think out - or model - all the implications of the rival
strategies.

If this wider consideration, and the evidence that could
be adduced from military manoeuvres and recent wars, did support
the basic case for dispersion, the new rule of thumb should be
"Disperse. Start from uniform dispersion and then ask
yourself how far, if at all, you wish to deviate from it".

4.   THE N-SQUARE LAW

Lanchester's Square Law can be addressed at two levels.
At the micro level, meaning in application to a single
engagement, we have seen from our numerical example how it works
if the assumptions are fulfilled, including in particular the
assumptions of low accuracy and no advantage to the defence.
At the macro level, meaning over a series of engagements
(either in the course of battle or a campaign) the validity of
the law is less obvious. An incompetent general facing a
competent general might enjoy overall numerical superiority and
yet never deploy his forces so that in combat they enjoyed
numerical superiority. For the law to be valid at the macro
level, the ratio of total forces has to be represented at the
micro level. If there are a large number of small engagements,
the requirement is that the ratio of total forces at the
average engagement should be that at the macro level - and some
dispersion around that average presumably will not matter.
If there are a few large engagements presumably less dispersion
of the ratio is acceptable.

If we bring in high accuracy and the consequent case for
dispersion - leaving aside the vulnerability of concentrated
forces to area bombardment - that alone seems enough to undo
the square law. At the micro level the individual engagement
is of a different kind. The advantage is with the defence;
numerical superiority does not help disproportionately. At
the macro level, the greater the relative numerical strength
of the attacker, the further he can go into the defensive
network, but the relationship between his numbers (given those
of his opponent) and the distance he can go seems to be linear.
I can see no square law here.

Whatever the success in the past of Lanchester's Square
Law in predicting the outcome of battles or providing a measure
of the relative strength of the armies of rival nations or
armies, there would seem to be increasing reason to doubt its
validity now. What will matter increasingly is who is on the
defensive and whether the attacker achieves surprise.

I have looked at some of the literature in which Lanchester's
Square Law has been used in recent years. One example is
Colonel Dupuy's historical study entitled "Numbers, Predictions
and War", [4]. He introduces, by historical judgement, so many

variables into his equations - he lists 73 of which 22 are used
in his "Operational Lethality Indices" and others are added to
allow for "combat circumstances" - that it cannot be said that
he is testing any law. As he admits, it is an exercise in
ad hoccery. If he shows anything, it is that no law fits.
What is perhaps interesting, however, is that the factors
he introduces for surprise become increasingly large in recent
years, in addition to which the author declares the effect of
surprise to have been greater in recent wars than before and
so multiplies the surprise factors for battles from 1966
onwards by 1.66.

A more recent example is the use of by Kaufman of the
Square law in an assessment of the relative strength of NATO
and the Warsaw Pact [5]. He reduces all ground forces to
firepower units (FPU's); he applies an adjustment factor for
fighting effectiveness, varying it according to whether the
forces are on the defensive or offensive; and he then applies
the square law to whole armies to see who wins. I am left
doubting the validity of this technique in which the advantage
of the defence is recognised but the square law is still used.
In the first place high accuracy may - though I don't know
when - invalidate the square law. Secondly I am troubled by
a technique in which all forces are reduced to common units
of firepower and a coefficient is then applied to allow for
whether they are used in offence or defence. The mix of
forces of either side may be better or worse suited to
defence, (eg. it may have different proportions of anti-tank
missiles to tanks or anti-aircraft missiles to aircraft), in
which case a general coefficient is inappropriate, specific
coefficients are needed, and the use of a simple law such as
Lanchester's becomes questionable.

But I have said enough. To sum up I can pose three
questions:

a) Does rising accuracy invalidate Lanchester?

b) Does that mean that the rule of thumb, or benchmark,
    from which the deployment of defensive forces should
    start is dispersion over the area to be defended?

c) Does it also mean that the square law should be
    abandoned and that the probable outcome of fighting
    should be assessed by pragmatic modelling?

Two appendices are attached. The first, by my colleague
Mr. Rowthorn, sets out an alternative mathematical formulation
of Lanchester's Law. The second notes some points about the
assumptions underlying the two formulations.

Appendix 1: A Simple Model of Artillery Duels by
Mr. R. Rowthorn

Consider a battle between two sides which takes place in
discrete "rounds". During any round each surviving gun fires
one shot at an enemy gun, chosen at random; each shot from
side 1 has a probability p_1 of hitting its target, and any gun
which is hit is put out of action completely. Let $n_i(t)$ be the
number of guns on side i when the t th round of firing commences.
Assume that $n_1(t)$ and $n_2(t)$ are "large". The probability of
any particular gun on side 2 surviving round t is given by:

$$(1 - \frac{p_1}{n_2(t)})^{n_1(t)} \tag{1}$$

Since $n_1(t)$ is large, the survival of any particular gun is
almost statistically independent of the survival of other guns.
Hence, by the law of large numbers, the proportion of weapons
surviving on side 2, (to take part in round t + 1), is given
approximately by:

$$\frac{n_2(t+1)}{n_2(t)} = (1 - \frac{p_1}{n_2(t)})^{n_1(t)} \tag{2}$$

Since $n_1(t)$ is large, this can be approximated by

$$\frac{n_2(t+1)}{n_2(t)} = \exp(-p_1 \cdot n_1(t)/n_2(t)) \tag{3}$$

By symmetry, it follows that:

$$\frac{n_1(t+1)}{n(t)} = \exp(-p_2 \cdot n_2(t)/n_1(t)) \tag{4}$$

Dividing (4) by (3), we get the following approximate formula:

$$\frac{n_1(t+1)}{n_2(t+1)} = \frac{n_1(t)}{n_2(t)} \cdot \exp(p_1 \cdot \frac{n_1(t)}{n_2(t)} - p_2 \cdot \frac{n_2(t)}{n_1(t)}) \tag{5}$$

This approximation is extremely accurate for large $n_1(t)$ and $n_2(t)$.

The balance of forces will improve in favour of side 1 during round t, if and only if:

$$\frac{n_1(t+1)}{n_2(t+1)} > \frac{n_1(t)}{n_2(t)} \tag{6}$$

which from (5) is equivalent to the following condition:

$$\exp\left(p_1 \frac{n_1(t)}{n_2(t)} - p_2 \frac{n_2(t)}{n_1(t)}\right) > 1 \tag{7}$$

which is in turn equivalent to:

$$\frac{p_1}{p_2} > \left[\frac{n_2(t)}{n_1(t)}\right]^2 \tag{8}$$

There is obviously an unstable process at work here. When large armies are facing each other, the side with an initial advantage will rapidly gain the upper hand. To see this, suppose the following condition is satisfied:

$$\frac{p_1}{p_2} > \left[\frac{n_2(1)}{n_1(t)}\right]^2 \tag{9}$$

From (8) this implies that the balance will shift in favour of side 1 during the first round, ie.

$$\frac{n_1(2)}{n_2(2)} > \frac{n_1(1)}{n_2(1)} \tag{10}$$

From (9) and (10) it follows immediately that,

$$\frac{p_1}{p_2} > \left[\frac{n_2(2)}{n_1(2)}\right] \tag{11}$$

and so during the second round, the balance will shift still further in favour of side 1, ie.

$$\frac{n_1(3)}{n_2(3)} > \frac{n_1(2)}{n_2(2)} \tag{12}$$

and so on indefinitely.

It is also clear from equation (5) that the balance will shift at an ever faster rate towards side 1 until eventually its preponderance is overwhelming.

Equation (8) can be interpreted as follows. To overcome a numerical inferiority on the other side by means of great accuracy is difficult. The relative accuracy required to offset numerical inferiority is equal to the square on the ratio of the larger arsenal to the smaller. Thus if side 2 has twice as many weapons as side 1 at the start of the battle, its weapons must have at least four times the kill-probability. This is Lanchester's Square Law. Note that this law only holds when the arsenals on both sides are large enough for the above approximations to apply.

Numerical examples

kill probability $p_1 = p_2 = 0.2$

	Side 1	Side 2	ratio
Initial position	2000	1000	2:1
After round 1	1810	670	2.7:1
After round 2	1681	390	4.3:1
After round 3	1605	165	9.7:1

kill probability $p_1 = 0.2$, $p_2 = 0.8$

	Side 1	Side 2	ratio
Initial position	2000	1000	2:1
After round 1	1341	670	2:1
After round 2	899	449	2:1
After round 3	602	301	2:1

Comparing the second example with the first, we see what a
large superiority in kill probability there must be to
compensate for the numerical inferiority of side 2.

Appendix 2: A Comment on the Two Formulations of
Lanchester's Law

I find Mr. Rowthorn's formulation, which shows round by
round what is happening in an exchange of fire, more
illuminating - if less economical - than Lanchester's
differential equation by reference to time. But there is one
point where, so far as I can see, both formulations depart
from reality for the sake of simplification. The point is
implicit in Lanchester's formulation, explicit in Rowthorn's.

Lanchester

In Lanchester's formulation he speaks not of a "hit
probability" or a "kill probability" when defining the
constants by which he multiples the strength of each side.
He speaks of the "fighting values of the units on each side"
(op cit p. 42). So far as I can see, his results make it
plain that he means kill probability. It is very clear (p. 51)
that he is referring to aimed fire, not area fire ie. to shots
aimed at individual enemy weapons or soldiers, not shots aimed
into an area where the enemy are known to be. For aimed fire
to have a constant kill probability per shot is odd. One
might expect the hit probability to be constant (reflecting
the skill of the soldiers). But you would expect that as one
side gained in superiority over the other its kill probability
would fall relative to its hit probability as more shots hit
targets that were already dead. That would not occur only on
two conditions.

The first is that the rate of fire is so slow and well
coordinated that shots are never fired at dead targets; and
more than one shot is never fired simultaneously at a live
target. But for these conditions to be satisfied at the end
of an unequal battle, when many survive on one side, aiming
at few men - and ultimately one man - on the losing side,
implies a very low rate of fire per man. But since Lanchester
assumes a constant rate of fire by reference to time, that
would imply slow firing at the start of the engagement.
Moreover to rule out that more than one man should aim
simultaneously at one target is to prevent full advantage
being taken of numerical superiority: for several men to aim
at one target is one way of seizing that advantage.

The second possible condition is that the hit probability of the winning side rises as it gains superiority. That is conceivable since diminishing enemy fire lets you aim more accurately.

Rowthorn

In Rowthorn's formulation the numbers on each side are large, so that "the survival of any particular gun is almost statistically independent of the survival of other guns". Further he assumes that the choice of targets is random. The result is to produce the square law but with a rate of attrition lower than in Lanchester's formulation, as noted earlier.

General

Do these peculiarities of the assumptions in any significant degree nullify or qualify the Square Law? I think not. They are the kind of points one comes upon when scrutinising assumptions, and the point of doing that in the present context is not to fiddle with Square Law but to see if it remains valid at all.

REFERENCES

[1] Afheldt, H., "The necessity, preconditions and consequences of a no-first-use policy", No First Use, ed. F. Blackaby and others, SIPRI, Taylor and Francis, London, 1984. A fuller version in German is to be found in Afheldt, H., "Defensive Verteidigung", Rowoht, Reinbek - Hamburg, 1983.

[2] Boserup, A., "Deterrence and Defense", The Bulletin of the Atomic Scientists, Vol 37, No. 10, Dec. 1981, pp. 11-13.

[3] "Diminishing the Nuclear Threat: NATO's Defence and New Technology, British Atlantic Committee, London, Feb. 1984, pp. 30-33.

[4] Dupuy, Colonel T.N., "Numbers, Predictions and War: Using History to Evaluate Combat Factors and Predict the Outcome of Battles", MacDonald and Jane's, London, 1969.

[5] Kaufmann, W., "Non-Nuclear Deterrence", in Alliance Security: NATO and the No-First-Use Question", ed. John W. Steinbruner and Leon V. Sigal, Brookings Institution, Washington D.C. 1983, pp. 67-71 and 208-216.

[6] Lanchester, F.W., "Aircraft in Warfare, The Dawn of the Fourth Arm", Constable, London, 1916.

STOIC: A METHOD FOR OBTAINING APPROXIMATE SOLUTIONS
TO HETEROGENEOUS LANCHESTER MODELS

P.J. Haysman
(Royal Ordnance Future Systems Group, RMCS)

and

K. Wand
(Operational Research and Statistics Group, RMCS)

1. INTRODUCTION

The Lanchester Square Law model of combat is a well-known
attrition model which has often been applied in both
deterministic and stochastic forms to direct fire homogeneous
battles - that is, with only one weapon type on each side.
Conceptually straight-forward extensions enable similar models
to be developed for battles in which each side contains more
than one type of weapon, the so-called heterogeneous battle.
However, these extended models (unlike the homogeneous models)
are generally not amenable to analytic solution, and recourse
must be made to numerical methods or simulation to obtain
estimates of, for instance, battle casualties or probabilities
of success.

The disadvantages of being entirely dependent on such
methods to obtain solutions particularly in terms of time
taken to obtain reliable estimates, are well known. In
this paper, therefore, an alternative method of solution is
explored. The approach taken is to transform a heterogeneous
model with time-independent kill rates into an "equivalent"
homogeneous model, and then to solve this using well-known
methods.

The aim of this paper is to describe the derivation of a
transformation method, STOIC (an acronym for Stochastic
Integrated Combat) and to demonstrate how, by its use, known
analytical results for the homogeneous Lanchester model may be
used to give approximate analytic solutions to heterogeneous
models which do not have exact analytical solutions.

2. THE BASIC LANCHESTER MODELS AND THEIR SOLUTION

Before describing STOIC and its applications, it is desirable to review (briefly) the basic formulation of both homogeneous and heterogeneous Lanchester models in their deterministic and stochastic forms and to discuss methods of solution available.

2.1 *Homogeneous deterministic model*

This is the simplest expression of Lanchester's Square Law. Consider two opposing forces, BLUE and RED, starting a battle at time t=0 with B weapons of a single type on the BLUE side and R weapons of a single type on the RED side. Then the instantaneous attrition rates of the forces at some subsequent time t are:

$$\frac{-db}{dt} = \rho r \; , \; r \geqslant 0 \qquad\qquad (1a)$$

$$\frac{-dr}{dt} = \beta b \; , \; b \geqslant 0 \qquad\qquad (1b)$$

where b,r are the force strengths of BLUE and RED respectively at time t,

β is the average kill rate of a single BLUE weapon against RED (assumed constant).

ρ is the average kill rate of a single RED weapon against BLUE (assumed constant).

These equations yield the time independent solution:

$$\beta(B^2 - b^2) = \rho(R^2 - r^2) \qquad\qquad (2)$$

or the time dependent solutions:

$$b = B \cosh \; (t\sqrt{\beta\rho}) - R\sqrt{\frac{\rho}{\beta}} \; \sinh \; (t\sqrt{\beta\rho}) \qquad (3a)$$

$$r = R \cosh \; (t\sqrt{\beta\rho}) - B\sqrt{\frac{\beta}{\rho}} \; \sinh \; (t\sqrt{\beta\rho}) \qquad (3b)$$

The factor $E^2 = \beta/\rho$ is important in determining the outcome of the battle. Three initial battle conditions are of interest:

if $E^2 B^2 > R^2$ then BLUE will 'win', ie., the RED force will be annihilated after a finite time whilst Blue has survivors,

if $E^2 B^2 < R^2$ then RED will 'win' after a finite time,

and if $E^2 B^2 = R^2$ (Parity Condition) then the forces will reach mutual annihilation after an infinite time.

The value of E may thus be regarded, in this instance, as a measure of the effectiveness of BLUE against RED.

2.2 Heterogeneous deterministic models

If the forces are composed of more than one weapon type on each side, the model formulation can be extended to take account of the effects of weapons of each type firing at each type on the opposing side. The overall attrition suffered by each BLUE weapon type is calculated by summing the effects of fire from weapons of each RED weapon type. The kill rates are dependent on both firer and target weapon types. The model can be extended in one of two ways:

Linear Weights Model. The generalised equations for a heterogeneous battle can take the form:

$$\frac{-db_i}{dt} = \sum_{j=1}^{n} \rho_{ji}\, r_j \quad , \quad i = 1,\ldots,m \qquad (4a)$$

$$\frac{-dr_j}{dt} = \sum_{i=1}^{m} \beta_{ij}\, b_i \quad , \quad j = 1,\ldots,n \qquad (4b)$$

where m is the number of BLUE weapon types,
    n is the number of RED weapon types,
    b_i is the force size of BLUE weapon type i at time t,
    r_j is the force size of RED weapon type j at time t,
    β_{ij} is the average kill rate of BLUE weapon type i
        against RED weapon type j,
    ρ_{ji} is the average kill rate of RED weapon type j
        against BLUE weapon type i.

Model with Fire Allocation Rule. An important factor that
will affect the course of the battle is the spread of fire by
each weapon between weapons of the different types on the
opposing side. In the last model, the effects of fire
allocation are subsumed within the average kill rates. However,
the allocation of fire can also be modelled explicitly giving
the generalised equations:

$$\frac{-db_i}{dt} = \sum_{j=1}^{n} \rho_{ji}\, p_{ji}\, r_j \quad , \quad i = 1,\ldots,m \qquad (5a)$$

$$\frac{-dr_j}{dt} = \sum_{i=1}^{m} \beta_{ij}\, p'_{ij}\, b_i \quad , \quad j = 1,\ldots,n \qquad (5b)$$

where now β_{ij} is the average kill rate of BLUE weapons type
i against RED weapon type j when all fire is
directed at RED type j

ρ_{ji} is the average kill rate of RED weapon type j
against BLUE weapon type i when all fire is
directed at BLUE type i

p'_{ij} is the proportion of fire from BLUE weapon of
type i directed against RED weapons of type j

p_{ji} is the proportion of fire from RED weapons of
type j directed against BLUE weapons of type i.

According to the assumptions made about the fire allocation
rules (represented by the value given to p_{ji} and p'_{ij}) the
model takes on a number of special forms of which two will
be referred to later:

(a)   Uniform Fire Allocation Model:

In this case each weapon fires indiscriminately at the
opposition. Therefore the proportion of fire directed against
weapons of a particular type will be given by the ratio of the
number of weapons of that type on the total number of targets:

$$p_{ji} = b_i \Big/ \sum_{k=1}^{m} b_k$$

$$p'_{ij} = r_j / \sum_{\ell=1}^{n} r_\ell$$

(b)   Parallel Duels Model:

Here, the battle is assumed to consist of a number of one-vs-one duels. The total number of such duels at any time will be the minimum of the number of weapons surviving on each side.

Hence,

$$p_{ji} = b_i / \text{Min} \left\{ \sum_{k=1}^{m} b_k , \sum_{\ell=1}^{n} r_\ell \right\}$$

$$p'_{ij} = r_j / \text{Min} \left\{ \sum_{k=1}^{m} b_k , \sum_{\ell=1}^{n} r_\ell \right\}$$

The solution of these models in general necessitates the use of numerical methods; one exception is the degenerate case in which one side has a single type of weapon only; an analytical solution is then possible.

Theoretically, an analytical solution to the Linear Weights Model is also possible, but is undoubtedly so complicated as to be useless for practical applications.

2.3 Homogeneous stochastic models

The homogeneous model described earlier may be given a stochastic formulation (see, for instance, [1]). A solution by simulation may be obtained for this by sampling from the distribution of the time between battlefield kills. This distribution is negative exponential with a mean time to next kill given by:

$$\frac{1}{\beta b + \rho r}$$

where b and r are the force sizes and β and ρ the kill rates.

The probability that, when a kill occurs, a BLUE weapon is
killed is:

$$\frac{\rho r}{\beta b + \rho r} \tag{6a}$$

and the probability that a RED weapon is killed is:

$$\frac{\beta b}{\beta b + \rho r} \tag{6b}$$

An alternative method for estimating the outcomes of
homogeneous stochastic square-law battles is to use a
recurrence method based on the relation derived by Gye
and Lewis [2].

$$R\, P_{B-1,R,y} + \frac{\beta}{\rho}\, B\, P_{B,R-1,y} = (R + \frac{\beta}{\rho}\, B)\, P_{B,R,y} \tag{7}$$

Initial conditions are $P_{0,R,y} = 0$ for $y \geqslant 1$ and $P_{B,0,y} = 1$
for $B \geqslant y$, where B and R are the initial force sizes of BLUE
and RED respectively; and $P_{B,R,y}$ is the probability that BLUE
will have at least y survivors when RED has been annihilated
(r=0). The overall probability of a BLUE win is obtained
by making y=1.

It is possible to use this method to calculate the approximate
probability of BLUE having a force size of at least y remaining
at the time when RED has a force size r=x (instead of zero as
above). Equation 2 gives

$$\beta(B^2 - b^2) = \rho(R^2 - r^2)$$

which may be scaled to an equivalent annihilation battle:

$$\beta(B^2 - b^2) = \rho(R'^2 - 0^2) \tag{8}$$

where $R'^2 = R^2 - x^2$.

Thus Equation (7) can now be used as an approximation for
battles which do not continue to annihilate by substituting
the nearest integer to R' for R.

2.4 *Heterogeneous stochastic models*

No simple relationship equivalent to the Gye and Lewis recurrence formula for the homogeneous stochastic case exists for the stochastic formulation of the heterogeneous battle. Recourse to a simulation method based on a logical extension of the procedure for the homogeneous model is therefore necessary to obtain solutions; see for instance [2].

3. TRANSFORMATION FROM A HETEROGENEOUS BATTLE TO A HOMOGENEOUS BATTLE

A description will now be given of the STOIC method for transforming a heterogeneous battle with constant kill rates into a equivalent homogeneous battle with similar overall attrition characteristics. This method will be illustrated first using the transformation of the so-called Linear Weights heterogeneous battle to a homogeneous square-law battle as an example. Comparisons will then be made between battle outcomes obtained using this approximation and results from other methods. The application of the STOIC concepts to other models - the Uniform Fire-Allocation model and the Parallel Duels model - is treated subsequently and similar comparisons made.

3.1 *The STOIC transformation of a linear weights model*

The Linear Weights Model has the equations:

$$\frac{-db_i}{dt} = \sum_{j=1}^{n} \rho_{ji} \, r_j \ , \ i = 1,\ldots,m \tag{9a}$$

$$\frac{-dr_j}{dt} = \sum_{i=1}^{m} \beta_{ij} \, b_i \ , \ j = 1,\ldots,n \tag{9b}$$

It was noted above that deterministic solutions of the above model for non-extreme data yielded values of

$$\bar{E}^2 = (R^2 - r^2)/(B^2 - b^2)$$

which were approximately constant over the course of the battle

$$\text{(where } B = \sum_{i=1}^{m} B_i, \ R = \sum_{j=1}^{n} R_j, \ b = \sum_{i=1}^{m} b_i \text{ and } r = \sum_{j=1}^{n} r_j).$$

Thus it appeared that it should be possible to perform an analytical transformation of the heterogeneous battle to yield an approximately equivalent square law battle.

$$-\frac{db}{dt} = \bar{\rho} r \ , \ r \geqslant 0 \qquad\qquad (10a)$$

$$-\frac{dr}{dt} = \bar{\beta} b \ , \ b \geqslant 0 \qquad\qquad (10b)$$

for which $\bar{E}^2 (B^2 - b^2) = (R^2 - r^2)$ where $\bar{E}^2 = \bar{\beta}/\bar{\rho}.$ (11)

It has indeed been found that an analytical value for $\bar{E}^2$ (which will meet these conditions) can be calculated from knowledge of the initial force sizes and the kill rates. Details of this analytical transformation (which forms the basis of the STOIC method) are summarised here:

(a)   Each b_i from the heterogeneous equations can be found in terms of b_e where e is a specific value of i, and each r_j in terms of r_f where f is a specific value of j.

(b)   These relations then yield a single equation which relates b_e and r_f .

(c)   The transformation condition is determined by considering the battle state in which the BLUE force is annihilated at exactly the same time as the RED force. This state is inserted in equation (11) yielding the expression

$$\bar{\beta}_{ef} \ B^2 = \bar{\rho}_{ef} \ R^2 \qquad\qquad (12)$$

where $\bar{\beta}_{ef}$ and $\bar{\rho}_{ef}$ are given by

$$\bar{\beta}_{ef} = \sum_{i=1}^{m} \frac{\beta_{if}}{C_i^e} \frac{B_i^2}{B^2} \tag{13a}$$

$$\bar{\rho}_{ef} = \sum_{j=1}^{n} - \frac{\rho_{je}}{C_j^{\cdot f}} \frac{R_j^2}{R^2} \tag{13b}$$

where $C_i^e = \dfrac{\displaystyle\sum_{j=1}^{n} \rho_{ji} R_j}{\displaystyle\sum_{j=1}^{n} \rho_{je} R_j}$ and $C_j^{\cdot f} = \dfrac{\displaystyle\sum_{i=1}^{m} \beta_{ij} B_i}{\displaystyle\sum_{i=1}^{m} \beta_{if} B_i}$ (14a and b)

(d) Finally all values of $\bar{\beta}_{ef}$ and $\bar{\rho}_{ef}$ are combined to give

$$\bar{\beta} = \sum_{e=1}^{m} \sum_{f=1}^{n} \bar{\beta}_{ef} = \sum_{e=1}^{m} \sum_{f=1}^{n} \sum_{i=1}^{m} \frac{\beta_{if}}{C_i^e} \frac{B_i^2}{B^2} \tag{15a}$$

$$\bar{\rho} = \sum_{f=1}^{n} \sum_{e=1}^{m} \bar{\rho}_{ef} = \sum_{f=1}^{n} \sum_{e=1}^{m} \sum_{j=1}^{n} \frac{\rho_{je}}{C_j^{\cdot f}} \frac{R_j^2}{R^2} \tag{15b}$$

The value of $\bar{E}^2$ is given by $\bar{E}^2 = \bar{\beta}/\bar{\rho}$ (as in equation (11)).

(e) $\bar{\beta}, \bar{\rho}$ and $\bar{E}^2$ are then estimators for the parameters of the equivalent homogeneous model given by

$$\bar{E}^2(B^2 - b^2) = (R^2 - r^2), \quad \bar{E}^2 = \bar{\beta}/\bar{\rho} .$$

It should be noted that although the transformation is obtained on the assumption of mutual force annihilation, it can be shown that the above equation may be used to investigate outcomes of the heterogeneous battle whether at an annihilation point or for other intermediate battle states.

It should be noted that the derivation depends on the assumptions that:

$$\frac{\sum\limits_{j=1}^{n} \rho_{ji}\, r_j}{\sum\limits_{j=1}^{n} \rho_{je}\, r_j} \quad \text{and} \quad \frac{\sum\limits_{i=1}^{m} \beta_{ij}\, b_i}{\sum\limits_{i=1}^{m} \beta_{if}\, b_i} \qquad (16)$$

are constant over the course of the battle and equal to C_i^e and C_j^f respectively. These assumptions are found to be good for a wide range of realistic input data.

Some simplification of the formulae above is possible if it is assumed additionally that all weapon types on each side are annihilated simultaneously. The transformed kill rates are then given by:

$$\bar{\beta} = \sum\limits_{e=1}^{m} \sum\limits_{f=1}^{n} \sum\limits_{i=1}^{m} \beta_{if}\, \frac{B_e B_i}{B^2} \qquad (17a)$$

$$\bar{\rho} = \sum\limits_{f=1}^{n} \sum\limits_{e=1}^{m} \sum\limits_{j=1}^{n} \rho_{je}\, \frac{R_f R_j}{R^2} \qquad (17b)$$

Although the latter assumption is untrue for a Linear Weights formulation - all the weapon types are in fact not annihilated at exactly the same instant - the use of the simpler Equation 17 has been found in many cases to give estimates of attrition in the initial stages of battle which are as good as those obtained using the full STOIC procedure described above. However, it is with the full procedure that the remainder of this paper will be concerned; the limitations imposed by the extra assumption just discussed need further investigation.

3.2 Numerical example of the transformation of a linear weights battle

As an illustration of results obtained when STOIC is applied to the transformation of a Linear Weights model, and subsequent comparison with solutions obtained by other methods, the following data will be assumed for a battle with two types of weapon on each side:

Starting strengths

$B_1 = 80$ $\qquad$ $R_1 = 60$

$B_2 = 24$ $\qquad$ $R_2 = 18$

	RED			BLUE		
Kill rates	β	1	2	ρ	1	2
BLUE	1	0.5	0.2	RED 1	0.75	0.5
	2	1.0	0.5	2	1.0	0.25

The attrition suffered by individual weapon types has been estimated numerically for the appropriate deterministic Linear Weights formulation of this battle and is shown in Fig. 1. The total force strengths at any time - a straight count of numbers remaining on each side - is also shown. It will be noted that BLUE has a total of nearly 40 weapons remaining when RED is annihilated.

The relationship between $(R^2 - r^2)$ and $(B^2 - b^2)$ during the course of the battle - where the symbols refer to totals over all weapon types - is indeed linear to a very good approximation. This confirms the observation on which STOIC is based, that total attrition can be represented by a homogeneous square law battle. The slope of the resulting graph is shown in Fig. 2.

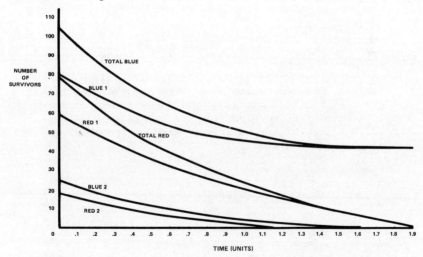

Fig. 1 Deterministic Result of linear weights Example

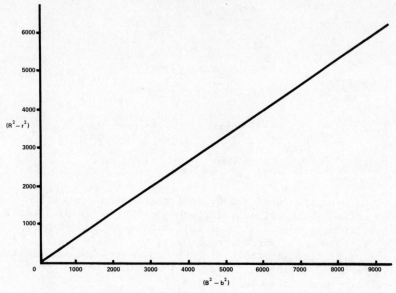

Fig. 2 $(R^2 - r^2)$ versus $(B^2 - b^2)$ for the linear
weights example.

Application of the STOIC transformation produces an
"equivalent" homogeneous battle with

$$B = 104 \qquad \bar{\beta} = 0.867$$
$$R = 78 \qquad \bar{\rho} = 1.282$$
$$\bar{E}^2 = \bar{\beta}/\bar{\rho} = 0.676$$

Table 1 compares estimates of BLUE survivors at two
different battle "end points":   first, at the point at which
the total number of survivors on the two sides together
is 150 (out for 182 initially); second, at the point at
which RED is annihilated.

TABLE 1: LINEAR WEIGHTS BATTLE EXAMPLE: COMPARISON OF OUTCOMES
ESTIMATED USING STOIC AND OTHER METHODS

SERIAL	Basis of estimate	Total (or mean) number of BLUE survivors	
		When RED and BLUE survivors total 150	When RED is annihilated
(a)	(b)	(c)	(d)
1	Heterogeneous deterministic	87.42	39.23
2	Heterogeneous stochastic	87.56 $\pm$ 0.05	42.20 $\pm$ 0.80
3	STOIC "equivalent homogeneous" deterministic	86.97	42.64
4	STOIC "equivalent homogeneous" stochastic	87.53	44.61

NOTE:

i)   If the extra assumption of simultaneous annihilation is
     made, estimates of BLUE survivors according to Serials 3
     and 4 become 87.4 and 88.1 respectively if RED and BLUE
     survivors total 150, and 47.1 and 47.4 when RED is
     annihilated.

ii) Serial 2 is based on 4000 replications (col. c) and
     400 (col d).

     The basis of these estimates is as follows:
     (a)   Serial 1. The battle data are inserted in the
           heterogeneous formulation of Equations 19 and
           solved deterministically by numerical methods.
     (b)   Serial 2. As Serial 1, but the equations are treated
           stochastically and estimates obtained from a simulation.

(c)   Serial 3. The "equivalent" homogeneous battle as
      obtained using STOIC is solved deterministically
      using analytical methods.

(d)   Serial 4. As Serial 3, but the homogeneous battle
      is now regarded stochastically and estimates made
      using the Gye-Lewis recurrence formula (Eq. 7).

It would be unwise to draw any general conclusions about
how satisfactory the STOIC transformation may be on the basis
of the one example given above. All that can be said here is
that a reasonable number of other comparisons that have been
made during the course of work at RMCS all lead to the
conclusion that the results of Table 1 may be regarded as
a typical illustration of the closeness of results obtained.
More work needs to be done however before the strengths and
limitations of STOIC can be fully realised.

One comment may be of interest. Perhaps the most
surprising aspect of the results shown in Table 1 is the way
in which the homogeneous battle obtained from the STOIC
transformation and treated stochastically (Serial 4 of
Table 1) produces estimates in good agreement with those
produced by simulation of the heterogeneous battle (Serial
2). This is further exemplified in Fig. 3 which shows the
distributions of BLUE survivors obtained by these two methods.
The value of STOIC as a practical tool will obviously be
enhanced if this agreement can be substantiated over a wide
range of realistic input conditions.

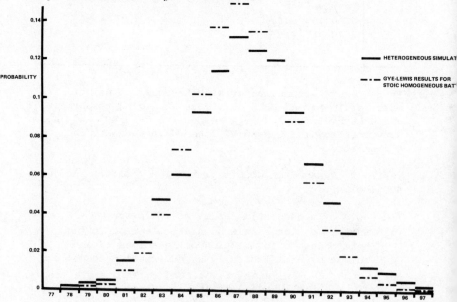

Fig. 3.   Distribution of Blue survivors when total survivors
          equal 150.

3.3 The STOIC transformation of a uniform fire allocation model

A second example of the STOIC transformation will use the Uniform Fire Allocation model, with the equations

$$-\frac{db_i}{dt} = \sum_{j=1}^{n} \rho_{ji} \frac{b_i}{\sum_{k=1}^{m} b_k} r_j, \quad i = 1,\ldots,m \qquad (18a)$$

$$-\frac{dr_j}{dt} = \sum_{i=1}^{m} \beta_{ij} \frac{r_j}{\sum_{\ell=1}^{n} r_\ell} b_i, \quad j = 1,\ldots,n \qquad (18b)$$

The STOIC transformation of this heterogeneous battle (see 2.2a above) to a homogeneous battle, was first considered in [3] and follows the same general method as that used with the Linear Weights model. The kill rates for the equivalent homogeneous square law battle in this case are found to be:

$$\bar{\beta} = \sum_{e=1}^{m} \sum_{f=1}^{n} \bar{\beta}_{ef} \frac{B_e R_f}{BR} \qquad (19a)$$

$$\bar{\rho} = \sum_{e=1}^{m} \sum_{f=1}^{n} \bar{\rho}_{ef} \frac{B_e R_f}{BR} \qquad (19b)$$

where

$$\bar{\beta}_{ef} = \sum_{i=1}^{m} \sum_{f=1}^{m} \frac{(\beta_{if} + \beta_{sf}) B_i B_s}{(c_i^e + c_s^e) B^2} \qquad (20a)$$

similarly for $\bar{\rho}$

$$\bar{\rho}_{ef} = \sum_{j=1}^{n} \sum_{t=1}^{n} \frac{(\rho_{je} + \rho_{te}) R_j R_t}{(c_j^f + c_t^f) R^2} \qquad (20b)$$

and c_i^e and c_j^f are as before (Eq. 14).

The use of this transformation will now be tested using the following example of a battle with three types of weapon on the BLUE side and two on the RED:

Starting $B_1 = 20$ $R_1 = 14$
strengths $B_2 = 10$ $R_2 = 4$
            $B_3 = 5$

	RED					BLUE		
Kill Rates	β	1	2		ρ	1	2	3
	1	1.8	1.2		1	3.0	3.8	2.4
BLUE	2	1.3	0.8	RED	2	3.4	4.0	2.7
	3	2.9	2.6					

The deterministic solution of this heterogeneous battle is shown in Fig. 4 and the relationship between $(R^2 - r^2)$ and $(B^2 - b^2)$ - where the symbols refer to totals over all weapon types - is shown in Fig. 5. The gradient of this graph is approximately constant (and equal to 0.53) again serving to confirm the observation on which STOIC is based.

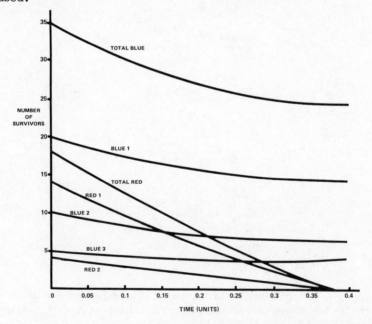

Fig. 4. Deterministic result of uniform fire allocation example

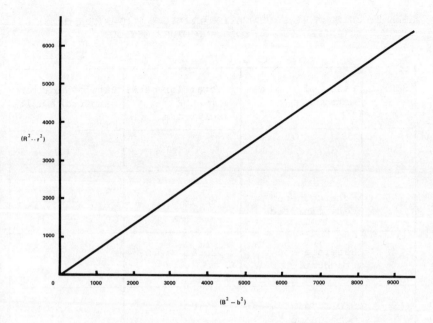

Fig. 5 $(R^2 - r^2)$ versus $(B^2 - b^2)$ for the uniform fire allocation example.

The "equivalent" homogeneous battle determined using the STOIC transformation has the following characteristics:

$$B = 35 \qquad \bar{\beta} = 1.748$$

$$R = 18 \qquad \bar{\rho} = 3.256$$

It will be noted that the estimated value of $\bar{E}^2 = \bar{\beta}/\bar{\rho} = 0.54$ agrees well with that obtained by the numerical solution of the heterogeneous equation and given by the gradient of the line in Fig. 5.

Table 2 gives results in a form similar to that used previously and demonstrates agreement similar to that already noted with the Linear Weights Model between estimates based on the STOIC transformation (Serials 3-5) and those made using more exact methods (Serials 1 and 2). The distributions of the number of BLUE survivors also conform as shown in Figure 6.

TABLE 2: UNIFORM FIRE ALLOCATION EXAMPLE: COMPARISON OF
OUTCOMES ESTIMATED USING STOIC AND OTHER METHODS

SERIAL	Basis of estimate	Total (or mean) of BLUE survivors	Battle duration (arbitrary units)
(a)	(b)	(c)	(d)
1	Heterogeneous deterministic	26.66	0.209
2	Heterogeneous stochastic	26.61 ± 0.07	
3	STOIC "equivalent homogeneous" deterministic	26.80	0.202
4	STOIC "equivalent homogeneous" stochastic	26.95	
5	STOIC "equivalent homogeneous" stochastic simulation	26.69 ± 0.064	

NOTE

i)   Estimates made as described for the Linear Weights
     Example. Serial 5 is an estimate based on a simulation
     method and compares with Serials 2 and 4.

ii) Survivors in column c are calculated for BLUE at a point
     when the total survivors on both sides is 34 (out of 53).

iii) Serial 2 is based on 2000 replications.

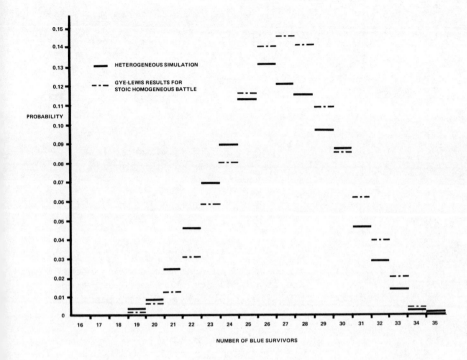

Fig. 6. Distribution of Blue survivors when total survivors
equal 34.

It will additionally be noted from Table 2 that estimated
battle duration based on both the exact and the STOIC
deterministic equations (Serials 1 and 3) are in good
agreement. It may be mentioned in passing that it is to
obtain the latter agreement that a weighted mean of all the
$\bar{\beta}_{ef}$ and $\bar{\rho}_{ef}$ is used in STOIC rather than any single pair
of values; although the ratio $\bar{\beta}_{ef}/\bar{\rho}_{ef}$ (for all e,f) and that
for a single $\bar{\beta}/\bar{\rho}$ may be very similar leading to similar
estimates of attrition, time scales between the two resulting
battles may be very different. The best agreement with the
exact solution occurs generally when a mean is used as
described.

3.4 The STOIC transformation of a parallel duels model

As discussed in 2.2(b) above, this model, representing
a heterogeneous battle fought as a series of parallel duels,
has the equations:

$$\frac{-db_i}{dt} = \sum_{j=1}^{n} \rho_{ji} r_j \frac{b_i}{\sum_{\ell=1}^{n} r_\ell} \quad , \quad i = 1, \ldots, m \qquad (21a)$$

$$\frac{-dr_j}{dt} = \sum_{i=1}^{m} \beta_{ij} b_i \frac{r_j}{\sum_{\ell=1}^{n} r_\ell} \quad , \quad j = 1, \ldots, n \qquad (21b)$$

(with $\sum_{i=1}^{m} b_i \geq \sum_{j=1}^{n} r_j$).

The STOIC transformation to an equivalent homogeneous battle follows the same lines and has the same assumptions as described previously. The kill rates of the transformed battle are:

$$\bar{\beta} = \sum_{e=1}^{m} \sum_{f=1}^{n} \bar{\beta}_{ef} \frac{B_e R_f}{BR} \qquad (22a)$$

$$\bar{\rho} = \sum_{e=1}^{m} \sum_{f=1}^{n} \bar{\rho}_{ef} \frac{B_e R_f}{BR} \qquad (22b)$$

where

$$\bar{\beta}_{ef} = \sum_{i=1}^{m} \frac{\beta_{if} B_i}{c_i^e B} \quad , \quad \bar{\rho}_{ef} = \sum_{j=1}^{n} \frac{\rho_{je} R_j}{\hat{c}_f^f R}$$

and c_i^e and $\hat{c}_j^f$ are as before (Eqn. 14). The homogeneous battle then follows a linear law

$$\bar{\beta}(B - b) = \bar{\rho}(R - r) \qquad (23)$$

As an example of the use of the STOIC transformation for this form of model, the same battle will be considered as for the Uniform Fire Allocation model. Figure 7 shows the

deterministic solution of the appropriate heterogeneous
equations, and Figure 8 the graph of $(R - r)$ against
$(B - b)$ where the symbols refer to totals over all weapon types.
It will be noted that this graph is linear with a constant
gradient equal to 0.546 confirming that the battle can be
transformed to an equivalent linear law battle:

$$\bar{E} (B - b) = (R - r)$$

where

(24)

$$\bar{E} = \bar{\beta}/\bar{\rho} \ (= 0.546).$$

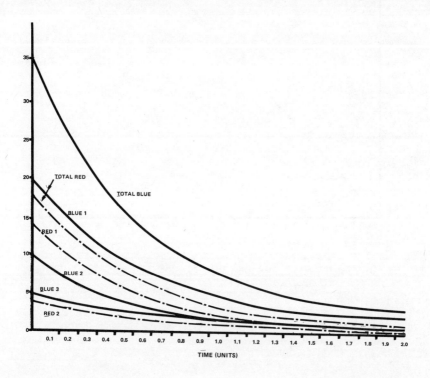

Fig. 7. Deterministic result for heterogeneous duels
example.

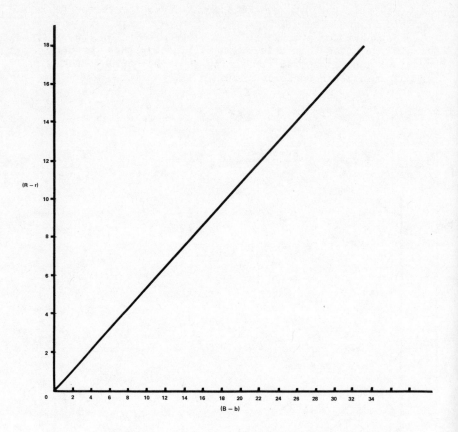

Fig. 8. (R - r) versus (B - b) for the parallel duels
        example.

The STOIC transformation yields:

$$B = 35 \qquad \bar{\beta} = 1.80$$

$$R = 18 \qquad \bar{\rho} = 3.30$$

This $\bar{\beta}/\bar{\rho}$ = 0.545 which is close to the value of $\bar{E}$ given
above.

TABLE 3: PARALLEL DUELS BATTLE EXAMPLE: COMPARISON OF
OUTCOMES ESTIMATED USING STOIC AND OTHER METHODS

SERIAL	Basis of estimate	Total (or mean) of BLUE survivors	Battle duration (arbitrary units)
(a)	(b)	(c)	(d)
1	Heterogeneous deterministic	22.57	0.137
2	Heterogeneous stochastic	22.56 $\pm$ 0.07	
3	STOIC 'equivalent homogeneous' deterministic	22.65	0.132
4	STOIC 'equivalent homogeneous' stochastic	22.71 $\pm$ 0.07	
5	STOIC 'equivalent homogeneous' stochastic analytical solution	22.71	

NOTE

i) Serials 1 - 4 correspond to estimates made as described
    for the Linear Weights Example. Serial 5 uses Equation
    (25).

ii) Survivors in column c are calculated for BLUE at a
    point when the total survivors on both sides are 34
    (out of 53).

Comparisons between estimates using the STOIC transformation
battles and more exact solutions are shown, as before, in

Table 3 and Fig. 9. The latter distribution as well as
Serial 5 of the Table are obtained by simulation and
analytically: the probability for the homogeneous stochastic
model that Blue will have y survivors when the total number
of survivors on both sides together is c is given by:

$$\binom{R+B-c}{B-y} \; p^{R+y-c} \; q^{B-y}, \text{ for } y = \text{Max } (1,c-R) \text{ to Min } (c,B) \quad (25)$$

where $p = \bar{E}/(1+\bar{E})$ and $q = 1 - p$.

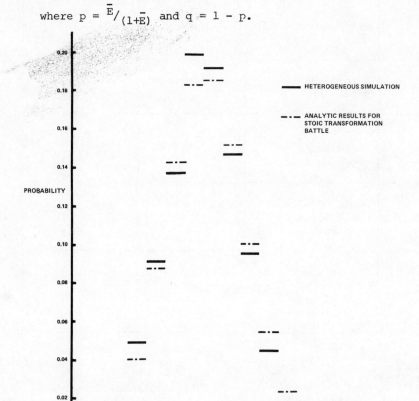

Fig. 9. Distribution of Blue survivors when total survivors
        equal 34.

Again agreement between estimates obtained by all methods for this example is good.

DISCUSSION

A description has been given of a transformation method, STOIC, which allows known methods of solution for homogeneous Lanchester battle models to be applied to otherwise intractable heterogeneous models in both the deterministic and stochastic models. Illustrations of the use of the transformation applied to heterogeneous battles of the so-called Linear Weights, Uniform Fire Allocation, and Parallel Duels configurations have demonstrated that the approximations in the method are reasonable in the examples quoted, and this conclusion is supported by other work not recorded here. However, it is too early to make any sweeping claims, and further investigations will be undertaken as time allows to obtain better insight into the generality of the method.

REFERENCES

1. RMCS, Report OR/C/18, STOCHADE: A Highly Aggregated Model of the Direct Fire Battle, Feb. 1980, Unclas.

2. Gye, R. and Lewis, T., Lanchester's Equations; Mathematics and the Art of War, 1976, Math. Scientist, Vol. 1, pp. 107-119.

3. Mortagy, B.E., Extensions to the Lanchester Model of Combat for the Analysis of Mixed Force Battles, 1981, PhD Thesis, RMCS.

THE MATHEMATICAL EQUIVALENCE OF CONCEPTUAL DIAGRAMS AND THE USE OF ELECTRONIC WORKSHEETS FOR DYNAMIC MODELS

G.J. Laing

(Defence Operational Analysis Establishment, West Byfleet)

1. INTRODUCTION

A model builder needs to set out very clearly all the basic features of any model he plans to construct. He also needs to be able to investigate the interaction of the variables over a wide range of values in order to develop a robust solution. It is hoped that the two techniques outlined in this paper will be helpful in attaining both these goals.

A first step frequently taken by operational research workers when attempting to analyse problems, is the construction of "conceptual diagrams". A conceptual diagram indicates which variables affect other variables although it does not give a precise mathematical description of the interactions between these variables. Such a diagram is also sometimes called an "interaction" diagram, a "cause and effect" diagram, a "flow" diagram, a "framework" diagram, or a "structure" diagram, since it helps to display the general structure of the problem. Graduates, fresh from University, are usually able to construct these diagrams without too much difficulty. It is the translation of the diagrams into working models that tends to cause more serious problems.

It is often not appreciated that, when drawn properly, conceptual diagrams outline the crude mathematical form of the problem and hence represent a very basic analysis of the situation. In this paper, some ways of developing more formal mathematical models from such diagrams are outlined. It is also suggested that by suitably rearranging the diagrams, they can readily be converted into dynamic models with the aid of standard "electronic worksheet" software packages. This permits some of the classical systems dynamics problems to be tackled without the use of sophisticated simulation packages, and also allows the rules used by decision makers to be more directly applied.

2.   CONSTRUCTING CONCEPTUAL DIAGRAMS

The first step which is usually taken in drawing conceptual diagrams is to set out the variables on a sheet of paper with lines showing which variable has a direct effect on another. Consideration can then be given to the coefficients and the stochastic effects. These parameters must be introduced with special care if the diagram is to be structured properly. It is a matter of personal choice, but I find the diagram easier to understand if I set out the variables in rectangular boxes, and the constants, constraints, coefficients and probabilities within circles. Thus the diagram for the Lanchester square law may appear as illustrated in Figure 1. In this case it will be recalled that:-

$$\frac{db}{dt} = - \rho r$$

$$\frac{dr}{dt} = - \beta b$$

where r is the size of the red force, b is the size of the blue force at a given time t, and β and ρ are the blue to red and the blue on red attrition coefficients (essentially, the "fighting value" of the individual blue and red units).

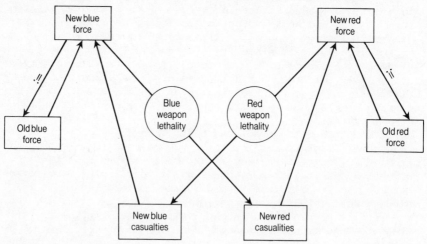

Variables are shown within rectangles.
Coefficients are shown encircled

Fig. 1.   A Diagrammatic Representation of the Lanchester
         Square Law.

The interesting feature about well laid out conceptual diagrams is that variables which are contained in the same chain of the diagram are multiplicative or divisive, but when several enter at a single nodal point they are additive or subtractive. However, when several variables enter via a coefficient they are also multiplicative or divisive.

The actions in this diagram are iterative, so only a portion can be considered at a time. Thus the centre diagonals form the righthand side of the Lanchester equations whilst the three boxes at the left and the three righthand side of the diagram relate to the rate of change aspects.

The Lanchester linear law provides a further example and this is shown in Figure 2. In this case we are concerned with situations such as "blanket fire" or the bombardment of a general area. It will be appreciated that in such situations the probability of hitting a target within the general aiming area diminishes as the battle continues.

The Lanchester formulae for this situation are as follows:-

$$\frac{dr}{dt} = -\beta br$$

$$\text{and } \frac{db}{dt} = -\rho rb$$

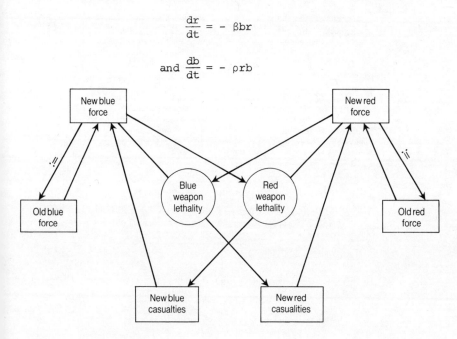

Variables are shown within rectangles.
Coefficients are shown encircled

Fig. 2   A Diagrammatic Representation of the Lanchester Linear Law.

Note that not only do the numbers of Red affect the number
of Blue casualties but also do the number of Blue. This is
due to the situation which often exists with blanket fire so
that when, for example, half the Blue Force is obliterated the
chance of Red hitting a Blue target is halved. It will be seen
that the two branches in the diagram leading into the α
coefficient, and the two leading into the β coefficient are
multiplicative in this case since they lead to a probability
coefficient (and are not additive or subtractive).

3. FROM DIAGRAM TO MATHEMATICAL MODEL

It is perhaps easier to explain the modelling method by
using a well known formula such as the Wilson formula for
studying an inventory problem. The normal mathematical
treatment leads to a formula for the annual cost of stocking a
certain line, given certain parameters such as the demand rate,
reorder quantity levels, and the costs of reordering, inventory
and of the item itself. The derivation of this formula is
given in most books on operational research or inventory and
stock control, and so it is not reproduced here.

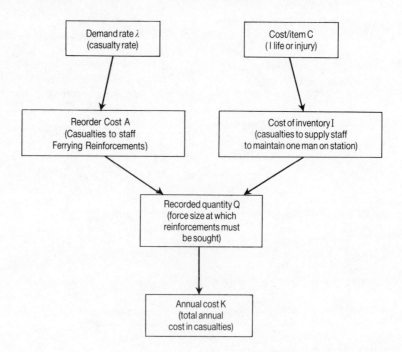

Fig. 3 A conceptual diagram of a military equivalent of the
        Wilson lot size inventory problem.

In practice this situation can have various military analogies. For example, it has certain similarities to those cases where a beleaguered garrison had to be maintained at Tobruk or Arnhem in World War II. Let us begin by first drawing the conceptual diagram showing which of the variables influence the others. The analogies to the military situation are shown in brackets underneath.

In this case the cost is measured in terms of human casualties and lives. The reorder cost includes the cost of the lives of airmen or seamen bringing reinforcements, and perhaps even the 'lives' (since they become involved) of the workmen rebuilding replacement planes and ships. The cost of inventory is the cost in terms of casualties to the support and supply teams of maintaining one fighting man in the garrison. Thus, finally, the total annual cost is the real cost in terms of manpower of maintaining the garrison for one year. The various steps are as follows:-

1. Find the box to which all the arrows lead. This is usually the objective of the formulation. Write this down on the left hand side of the page (it will become the left hand side of an equation).

2. Find all the separate branches leading from this and write these down separately on the right hand side. A plus or minus sign will occur in the formula to join each branch. Thus:

$$K \quad \Big| \quad \pm (\lambda, A, Q) \pm (C, I, Q)$$

At this stage it would be possible to write the formula as:

$$K = \pm(a\lambda^g.bA^h.cQ^k) \pm (dC.^\ell eI^m.fQ^n)$$

where a b c d e f g h k ℓ m and n are constants.

3. Now consider the left-hand branch of the diagram. Does this increase or decrease the value of the left hand side of the equation? If the answer is 'Yes' this term is positive. Then take the branch to the right hand side of the one just considered and so on until all the branches have been treated. The formula can now be written:

$$K = (a\lambda^g.bA^h.cQ^k) + (dC^\ell.eI^m.fQ^n)$$

The following procedures indicate whether the constants
g h k l m n are positive:

4. Start at the end of the left-hand branch. If this
variable is increased will the left hand side of the
equation increase or decrease? If it increases place
the variable in the numerator otherwise place it in
the denominator (i.e. the power is negative).

5. Progress to the next variable on the branch and
consider whether this increases or decreases the left
hand side having regard only to the preceding variables
in this branch.

In other words in the example, from Rule 4 above, the term
$a\lambda^g$ will appear in the numerator. Taking the next variable
in the branch, as per Rule 5, it will be seen that bA^h will
also appear in the numerator. Now the effect of reorder
quantity Q on this branch can be seen and it will be noted that
the cost K on the left hand side will be increased by
decreasing Q since the reorder costs will be more. Hence, in
this case the Q term appears in the denominator.

Thus this branch can be written:

$$\frac{a\lambda^g \cdot bA^h}{cQ^k}$$

where the K is now positive in the formula.

Similarly by developing the other branch it is possible to
write the more complete formula:

$$K = \frac{a\lambda^g \cdot bA^h}{cQ^k} + dC^\ell \cdot eI^m \cdot fQ^n$$

Now consider the values of the coefficients a b c d e and f.
This evaluation of constants may or may not be possible by
argument. The following forms the basis of this decision.

6. Is the left hand side likely to change with the
weighting of the variable? If the constant is doubled
and other branches are removed the left hand side will
be doubled. Does this seem right? Should the term be
unity or is there some reason for another value?

In the example this form of argument shows that a b c d and e should be unity but f is equal to 0.5 since the cost K on the left hand side will depend on the average stock. Thus the formula can now be written:

$$K = \frac{\lambda^g A^h}{Q^k} + c^\ell I^m \left(\frac{Q}{2}\right)^n$$

Finally consider the power of the variables as follows:

7. How will the left hand side be likely to change with power of the variable? If the other branches are removed and the variable to be considered is changed in value will the power constant induce the correct changes in the left hand side? Should there be any other changes?

This part of the problem is often rather difficult and computation and statistics may be necessary for the evaluation of both the coefficients and the power terms. The latter may be especially difficult if they cannot be visualised as whole numbers. The solution to this example is relatively simple and after some thought it will probably be appreciated that in this case these terms are likely to be unity. Thus the final answer can be expressed as:

$$K = \frac{\lambda A}{Q} + I\ C\ \frac{Q}{2}$$

This is precisely the solution obtained by using the normal mathematical approach, although it must be admitted that in this case the process has been a great deal more lengthy. However, the object here has been to illustrate the method as clearly as possible rather than to obtain a rapid solution. It is hoped that this approach will enable the step from conceptual diagrams to formulae to be made more easily in quite complicated situations. Some thought is necessary in constructing the diagrams, but if the constants and coefficients are initially sorted out very carefully and related correctly to their affective variables in the diagram, the work may be appreciably reduced.

4. USE OF ELECTRONIC WORKSHEETS

It is, of course, possible to use conceptual diagrams in conjunction with statistical analyses. Methods such as regression, component analysis, factor analysis, cannonical correlation analysis, discriminant analysis and cluster analysis may help in deriving both the shape of the diagram and the coefficients. Some of these techniques have come to be employed to a much greater extent in recent years as a

consequence of the increasing use of computers since the latter
save a great deal of time and drudgery in executing the
necessary calculations. Unfortunately these methods presuppose
the existence of large quantities of data and since there are
few modern battles to provide this kind of information they are
often of limited value for military studies.

Even so, in many military situations we have some
understanding of the mathematical relationships between the
variables and may simply wish to investigate the effects of
changing constraints, coefficients, or initial values. There
is an interesting way in which models for this purpose can be
constructed with the help of conceptual diagrams and spreadsheet
software. It is, in fact, normally essential to reconstruct
the diagram in a new form so that it can ultimately be used more
directly in conjunction with certain standard software packages.
It is first necessary to establish the relationship between the
variables, and also the precise sequencing mechanism, but having
done so it is immediately possible to enter the mathematical
relationships directly into a matrix at the appropriate points.
In order to do this, one has to visualise the whole empty
matrix as being sequenced in some way (usually time) either
sideways or in an upwards or downwards direction.

Once the problem has been reformulated in this manner it is
possible to transfer it to a microcomputer which uses an
electronic worksheet package such as "Visicalc". This kind of
package was primarily designed for accounting and budgeting
activities, and as a consequence, the originators do not seem
to be aware of its full potential. A Visicalc worksheet
consists of 63 columns and 254 rows. Each cell can contain an
individual formula precisely describing how the value in the
cell is calculated. This formula is displayed on the visual
display unit by moving a cursor onto the particular cell under
review. As the cursor is moved to the next cell, the value
of the old cell is calculated and entered into the matrix.
Formulae can be replicated in rows and columns relative to
other values, and the whole matrix calculated automatically.

It only takes a few minutes to build what is known as a one
by one battle (one category of weapon on each side) using this
method. The matrix can be laid out to time sequence in either
direction, depending on the number of time sequences or weapon
categories required. A one by one battle would obviously be
laid out with the different surviving forces, casualties, and
lethality coefficients, separately in columns. This would
give some 254 possible states for the variables and far more
than are necessary to represent the non-linear curves given by
a set of equations. (It should be noted that linear equations
do not have to be used anyway. They just help to keep the
model simple in this case).

Rectangles denote variables
Circles denote lethality coefficients

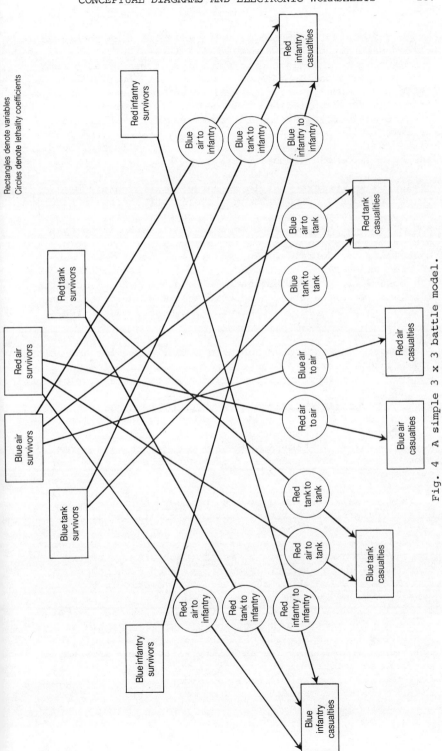

Fig. 4 A simple 3 x 3 battle model.

In this particular example the matrix was time sequenced
across each row and then downwards. It was then a simple
matter to make changes and to demonstrate the effect of one
side having twice the size of force and the other side having
twice the weapon power. The well known Lanchester square law
effect is demonstrated and the numerically superior side wins
despite the better weapon effectiveness of the other side. It
is immediately possible to go on to show that weapons four
times as good are needed to balance out the effect of doubling
the numbers of one of the combatants.

The three by three battle illustrated in Figure 4 was
constructed very quickly using Visicalc and it has a moderate
degree of complexity in that attrition coefficients change in
some cases as the variables change. This is achieved by a
series of 'IF statements' within the cells of the matrix
containing the formulae. The model took about twenty minutes
to plan and set up. Very much more complicated battle models
can be quickly produced using this method and it is possible
to investigate the effects of more than a hundred different
categories of weapon on either side, i.e. a "one hundred by
one hundred" battle.

Visicalc does not have a random number generator and so the
model cannot be made stochastic without additional effort.
However, a simple mathematical random number generator can be
included in the matrix and this can be arranged to change the
attrition coefficients after every run. It is possible to log
the outcome of all the runs quickly or send them to file. As
a consequence, one can carry out and log, say, twenty runs in as
many minutes for a moderate sized model. This will normally
provide a stochastic sample of adequate size for most
purposes.

In this way, the whole process of model building becomes
a relatively simple activity. Naturally, all the difficulties
normally encountered in building models are not entirely
removed. The task of selecting the important variables and
their relationships is not obviated. Nor are any of the
problems which can sometimes be experienced when other
variables, outside those currently incorporated in the model
system, suddenly assume importance. However, by carefully
using the process just outlined, together with the Visicalc
software, the model can be developed logically within a
prescribed framework which ensures that every variable is
considered individually as an entity without any programming
language distractions, or worries as to precisely where it
fits into the program, or added considerations as to whether
the additional program will in some way invalidate previous

programming, possibly by creating looping errors.

A wide variety of military models can be developed using this technique. For example, a simple reconnaissance model can be produced in which a number of reconnaissance aircraft are flown against different areas, there being certain probabilities of useful detection and certain probabilities of aircraft being shot down. These probabilities can be changed with time in order to allow for learning processes. The Visicalc page can be arranged so that the commander of the reconnaissance force is progressively able to view the left-hand portion of the page. The opposing commander must move his forces through various areas deploying his counter-aircraft support as he wishes. This can form the centre section of the Visicalc page and can be viewed separately so that the information of each commander is discreetly stored. The ground commander must plan his movements the requisite time ahead on this portion of the page. The value of the information available in each area may be modified by data acquired as a consequence of successful reconnaissance flights.

The right-hand part of the Visicalc page can simply be used as a work area to enable the program to run effectively. It can contain calculations to determine the cumulative probability of an aircraft being shot down over one of the reconnaissance zones and the cumulative probability of finding a target therein. The model may be deterministic, but it can be made stochastic by the inclusion of random number streams in the matrix.

5. MODELLING "SYSTEMS DYNAMICS" PROBLEMS

It is important to appreciate from the foregoing that spreadsheet packages can be used for those activities currently described as systems dynamics and for rapid demonstration of the performance of such systems. Consequently it was decided to carry out a brief study to see whether these methods could provide a reasonable alternative to the more normal system dynamic approaches using special simulation languages such as DYNAMO and DYSMAP. In many respects the models already described can be said to be "dynamic", but the question remained as to just how far a classical system dynamics solution could be emulated. Perhaps one of the most effective of Forrester's system dynamic models is his model of an industrial situation [2]. It was therefore decided to use this model for a comparative study rather than his urban dynamic model [3], or his model of world dynamic behaviour [4].

The classical problem set out by Forrester involved a
factory producing items which passed through a distributor
to a retailer, there also being various storage points in the
system. Coyle [1]   talks about "rates" and "levels".
"Rates" are the equations which govern the flows, and the
"levels" are stock levels, or levels attained by the variables
under consideration. Forrester applied step and ramp
functions to the retail demand levels and showed that this
could cause serious oscillations in the distributer's
supplies, in factory production, and in other variables.
Figure 5 shows the oscillatory effects produced in the system
by a sudden "stepped" rise in the demand.

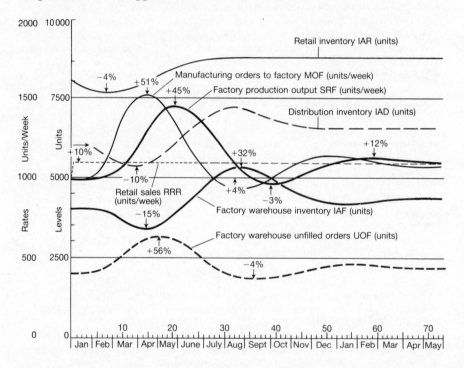

Fig. 5 Oscillatory effects demonstrated by Forrester as a
        consequence of a sudden 10 per cent rise in demand.

The first task in constructing the model was to sketch
out a rough chain of events in conceptual diagram form,
starting with the initial receipt of an order for a product.
These were then laid out in approximate sequence across the
top of the Visicalc page, and time delays between each
event were estimated. The numbers entered into the columns
beneath represent either a stock level or a rate of flow at

some point in the system. As explained earlier, each cell in
the Visicalc matrix can contain a small program which enables
the value at that point to be calculated. Since the steps in
this model are small, the interactions between the variables
are reasonably simple. This means that the program can be
accommodated within the cells and that virtually no further
columns need to be added exclusively for calculation
purposes. [5] .

It therefore follows that the decision rules which are
applied to each column correspond to the decisions which have
to be taken by a Departmental Manager or by a Section Head
within the Department. Hence, each column must be treated
separately and a mathematical expression, or program, devised
which corresponds to the decision rules of the department
being considered. Thus, for example, in the case of a
distributor deciding what orders to place on a manufacturer it
may be necessary for him to consider such things as his
current stock level, the demands he is receiving from
retailers and his desired stock level. After all, what
happens in practice is that policies are decided after a finite
time interval and decisions are based on all the affective
input variables at that time. Thus we tend to think of
placing orders at say weekly or monthly intervals. Some
would argue that this kind of discrete model may be more
realistic than a system dynamic model where continuous flow
rates have to be estimated from observation and prediction.
It is also rather more meaningful, for example, to say to a
manager of a purchasing department "What causes you to decide
to place a new order" rather than to say "What is your
anticipated dy/dt and why?"

Perhaps, and this may be even more important, the method
simplifies the task of model building. With this technique,
elaborate courses or training are not necessary and anyone
with a rudimentary knowledge of programming can learn to
build their own dynamic models within the space of a few
hours. An example of the output from the spreadsheet program
is given in Figure 6, when after seven time periods of
complete stability, the shop demand was changed by creating
a 16 per cent step function increase. The oscillatory
effects were still evident some forty one time periods after
this step function had been applied and the similarity of
the phrases and the oscillations with Forrester's results will
be evident. The linearity imposed on parts of the oscillations
shown at Figure 6 are due to the different purchasing policies
resulting from changed circumstances. It is suggested that in
a real world these are very much a part of the pattern of
events which take place.

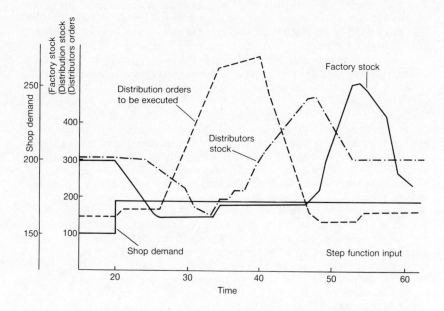

Fig. 6   An illustration of the effects created with a factory
         system dynamics problem of the type suggested by
         Forrester but in this case using spreadsheet software.

5.   CONCLUSIONS

Clearly there are many military situations where system
dynamics could be used [6] and where this method is easier
to apply than the current conventional approach. However,
each case must be decided on its own merits.

It is not suggested that the approaches outlined in this
paper will solve all problems, but rather that if the methods
which have been described are followed, a better understanding
will be obtained and hence in some cases a solution will be
more obvious.

The reformulation of conceptual diagrams in a time sequenced
mode together with the employment of "electronic worksheet"
type software packages provides a powerful means of constructing
models quickly. Moreover, in the industrial problem just
described, it did seem to take the analysis one stage nearer
the real world and to cause the program to be formulated more
directly from the statements made by the decision makers.

This latter method is not, of course, confined to the use of Visicalc and it seems likely that most electronic worksheet packages offering a reasonable matrix size could be adapted for this purpose. Thus Supercalc, Plannercalc, Masterplanner Calcstar etc., (all rather similar packages) could probably equally well be used in this way.

Finally, the ease and speed with which models can be constructed may mean that in some cases the decision makers will be able to formulate their own models rather than employ specialist help. It is likely that under the circumstances they will have a better understanding of the strengths and weaknesses of the models they are using.

REFERENCES

1. Coyle, R.G., (1977)   "Management Systems Dynamics", Wiley.

2. Forrester, J.W., (1961)   "Industrial Dynamics", MIT Press.

3. Forrester, J.W., (1969) "Urban Dynamics", Published by MIT Press.

4. Forrester, J.W., (1971)   "World Dynamics", Wright-Allan Press.

5. Laing, G.J., (1981 and 1982) Internal Ministry of Defence papers.

6. Laing, G.J., (1986) "Scientific Models", Academic Press.

LANCHESTER THEORY IN PRACTICE: A Constructive Critique of
the Mathematical Modelling of Combat in Defence
Operational Analysis

W.T. Lord
(Rex, Thompson & Partners, Farnham, Surrey)

1. INTRODUCTION

1.1 Aim

The analysis of military conflict is widely known in the
UK as defence operational analysis. A prominent feature of
defence operational analysis, or defence OA for short, is the
mathematical modelling of combat. Severe criticisms have been
voiced about the part played by the mathematical modelling of
combat in defence OA. My aim in this paper is to describe some
of the criticisms and to put forward some suggestions for
reducing their force.

1.2 Background

I have drawn my material mainly from the experience I gained
from 1974 to 1982 at the Defence Operational Analysis
Establishment (DOAE), West Byfleet. I entered defence OA with
a foundation of theoretical and experimental knowledge of
fluid dynamics. At DOAE I worked exclusively in the Studies
Divisions, taking part in various UK Studies about sea, land
and air operations, generally as Study leader. I was also
involved in several international meetings. I was not at any
time officially engaged in research at DOAE, but I acquired a
peripheral acquaintance with the more academic side of
defence OA.

In this paper I have tried to blend my theoretical knowledge
with my practical experience of Studies. Most of the points
were raised during my days at DOAE. The paper itself had its
origins in a talk given in October 1983 to the informal UK
Lanchester Study Group, as described by Daly [19] . The
argument was taken further in two formal presentations, the

first at the Shrivenham Symposium in September 1984 (Lord [61]),
and the second in December 1984 at the Cambridge Conference of
which this volume is the proceedings. In preparing this paper
from the text of the latter presentation (Lord [62]) I have
taken the opportunity to refer to some related papers published
in 1985.

The paper has a short title and a long sub-title. In fact,
the sub-title describes the subject matter more accurately
than the title, which is preserved for continuity with the talk.
The connection between the title and the sub-title is shown in
Figure 1. The meanings of the headings in Figure 1 will become
clearer as the terms are explained in the text.

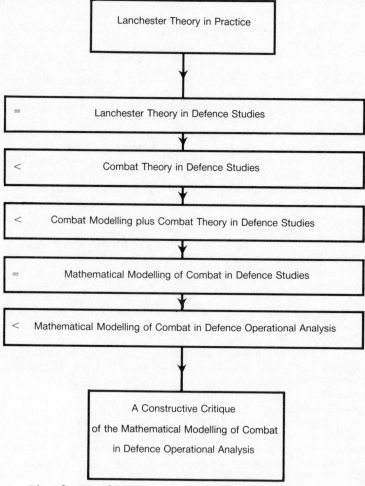

Fig. 1   Development of Sub-title from Title

The explanations of the terms are also given in the Glossary.
The need for a Glossary is not unusual in defence OA. My
explanations are not intended to be absolute definitions, to
be defended against all attacks. They are personal
descriptions, composed in the hope that they will help to
clarify what I am trying to say in this paper.

Few of the ideas in the paper are likely to be new
individually, and some are becoming increasingly well
established. However, in their totality they may not have been
put across in this way before. Taken together they make up a
scheme with room for recognisable compromises to be made at
certain stages in the use of the mathematical modelling of
combat in Studies. They may thus help to identify and isolate
the truly crucial difficulties of defence OA.

One thread in the argument is that there seems to be a blind
spot in defence OA where mathematics is concerned. It might
be thought that mathematics would be applied with telling effect
throughout defence OA. This is not so. In many cases defence
OA in the UK does not use mathematics in some of its most
elementary forms. Indeed, several of my suggested measures are
simple mathematical devices, standard in fluid dynamics for
instance, which are often neglected in defence OA.

To keep within the bounds of brevity set by the editor I
have omitted much material which would have been included in
a more detailed and more operationally-oriented article. On
the other hand, in line with a second point of editorial policy,
I have tried to make the paper intelligible to the non-
specialist by giving more survey information than might
otherwise have been needed. For the sake of compression I rely
on references to past work for expressions of criticism and for
examples of the application of some of the constructive
suggestions (there does not exist a comprehensive example).

The paper contains a list of one hundred references. Some
are marked as for background reading and are not mentioned
specifically in the text. The listed documents form only a
small fraction of the unclassified references available, and I
trust that in my selection, and in the paper as a whole, I have
not displayed more than my share of ignorance and bias. Most
of the more substantial works come from the USA. They include
the classical text by Morse and Kimball [67] and the remarkable
book by Dupuy [23]. Some of the works of recent origin, such
as that by Taylor [86] and those edited by Shubik [81] and
by Hughes [45], have greatly enhanced the open literature on
the mathematical modelling of combat. However, it should be
borne in mind that material free from any security classification
is merely a minute part of the total information that exists

on defence OA. The most potent sources of an authoritative
open article are sometimes hidden beneath the necessary mantle
of security.

1.3 Outline of Paper

The text consists in effect of two parts and a concluding
section. The first, general, part (Sections 2,3,4) begins with
a short survey of defence OA (Section 2). The survey is
followed by some critical comments on the role played by the
mathematical modelling of combat (Section 3). Section 4 then
gives a broad description of the constructive suggestions. The
second part (Sections 5,6,7) is more detailed. It describes
some features of Lanchester theory (Section 5) as an
introduction to the particular measures, given in Section 6,
for assisting the use of the mathematical modelling of combat
in Studies. Section 7 deals with work outside Studies. The
paper ends with a summary of the main points (Section 8).

Defence OA is not an easy subject, and partly for that
reason, the paper may not be easy going in places. Conceivably
non-specialists might make something of it; the casual reader
might well go straight to Section 8 and even the interested non-
specialist should probably omit Sections 5 and 6. Old-fashioned
applied mathematicians with an inclination towards research into
the problems of defence operations might get through the whole
paper. But it is directed principally at defence operational
analysts in the UK, as a contribution to the long-running
debate about the role of the mathematical modelling of combat
in defence OA. It may seem an odd paper in the age of
information technology and artificial intelligence, but one of
its themes, that there is a need for analytical understanding in
defence OA, should hold good in any age.

2. A SHORT SURVEY OF DEFENCE OA

2.1 Subject

The general subject of operational analysis (OA) grew out
of the military operational research undertaken in the Second
World War. In civil applications it is still called operational
research (OR). However, as Price [73] has recalled, its name
in military applications was changed in the 1960s to avoid
confusion with operational requirements when abbreviated. I
shall call it OA throughout this paper and take it to be
synonymous with operational research.

The shortest definition of OA that I know is "quantified
commonsense". While this description has a lot to recommend it
the traditional definition quoted by Morse and Kimball [67]

and adopted by Taylor [86] is more to my purpose. I shall
adapt it slightly and take OA to be an aid to decision-making
which introduces a quantitative element, otherwise absent, by
using a scientific method. Defence OA is then, naturally, the
branch of OA which deals with the problems of defence.

Consider the definition of OA more closely. It consists in
effect of two separate parts, one concerned with decision-
making and the other with quantitative analysis. While the
quantitative analysis may be described, at least loosely, as
science, the interpretation of the quantitative element as an
aid to decision-making is more of an art. Defence OA is
therefore an art as well as a science. The art produces the
analytical judgement which is the end-product of a piece of
operational analysis, while the science contributes to the
analytical experience on which that judgement is based. The
idea of the two distinct parts of defence OA is displayed in
Table 1a. It will be developed further in Table 1 and followed
throughout the paper.

Defence OA is a vast, highly complex and rather strange
subject. Its development in the UK seems to have been
constrained by the way it is organised. The operational and
analytical aspects of defence OA are, for the most part, only
brought together in specific Studies, where analytical
experience and analytical judgement intermingle. The
organisation does not provide for the steady building-up of a
permanent body of authentic quantitative information that might
be called the science of defence operations. To elaborate on
these statements I shall deal in turn with the operational
aspects, analytical aspects and organisation of defence OA,
with particular reference to DOAE.

2.2 Operational Aspects

Defence operations cover all the activities undertaken by
the Armed Services in peacetime, tension and war, including
nuclear, biological and chemical as well as conventional
warfare. They include all levels of operations at sea, on
land and in the air, from broad strategic concepts to detailed
tactical options in the use of weapon systems, sensors and
support equipment. The most prominent activity is combat.
The branch of defence operations concerned with combat is often
called combat dynamics, a useful phrase apparently coined by
Brackney [9] . The aspect of combat dynamics most commonly
considered is the attrition of the combatants.

Tables la-ld Identification of Four Principal Components of Analysis

TABLE 1a First Major Distinction, derived from Definition of Defence OA	Decision-Making	Quantitative Analysis		
	Art	Science		
	Analytical Judgement	Analytical Experience		
	Subjective	Objective		
	Soft Analysis	Hard Analysis		
	Saaty Method	Mathematical Modelling of Combat		
TABLE 1b Second Major Distinction, contained within Mathematical Modelling of Combat		Combat Modelling	Combat Theory	
		Simulation Modelling	Lanchester Theory	
TABLE 1c Third Major Distinction, based on Organisation of Defence OA	Practice of Defence Studies		Analysis of Defence Operations	
	Inside Studies		Outside Studies	
TABLE 1d Four Principal Components of Analysis	Soft Analysis Inside Studies	Combat Modelling Inside Studies	Combat Theory Inside Studies	Combat Theory Outside Studies
	Saaty Method Inside Studies	Simulation Modelling Inside Studies	Lanchester Theory Inside Studies	Lanchester Theory Outside Studies

The scale of operations in combat dynamics may vary from
that of a war as a whole down through a campaign, a battle and
an engagement to a duel. Combat is often of a fragmentary
nature. It depends on the particular scenario of time and
place, and on the characteristics of the weapon systems employed.
Sea-air operations involve many scenarios, relatively few weapon
systems. Land-air operations involve many weapon systems and
relatively few scenarios. Combat may take place between forces
of different sizes and compositions. It may be critically
influenced by logistics, the nature of the physical
environment and by the tactics adopted by the two sides. Of the
greatest importance are human factors, such as training and
morale, and the command, control, communications and intelligence
systems, which bring together, so far as allowed by the
intricacies of electronic warfare, the intentions and actions of
friendly and enemy forces.

Nowadays, defence OA is concerned not only with current
operations but also, perhaps to a greater extent, with future
planning. It deals with the problems of resource allocation to
meet possible threats and it has a developing role in military
training. The planning role includes giving assistance in
deciding between alternative military policies, for instance, or
between competing procurement options. Defence OA often
requires the estimation of the overall combat effectiveness of
a force in relation to its projected cost. The estimation of
cost (not considered in this paper) is as important as the
estimation of effectiveness in any cost-effectiveness analysis.

From now on I shall concentrate on combat dynamics and in
particular on combat effectiveness as analysed through the
attrition of opposing forces. The notion of quantifying combat
effectiveness dates back to 1905 and is credited to Admiral
Fiske of the United States Navy. Fiske, as reported by Weiss
[94], said that 'there is always some numerical factor that
expresses the comparative value of two contending forces, even
though we never know what that numerical factor is'. We still
do not know today what the numerical factor is, but trying
to find it is one of the most stimulating activities of
defence OA.

2.3 Analytical Aspects

Quantitative analytical experience is, or should be, gained
scientifically. It may be said to derive from 'hard' analysis.
The art of analytical judgement also has its techniques, but
they are subjective rather than objective, 'soft' rather than
hard. Henceforth I shall associate the decision-making part
of defence OA directly with soft analysis and the quantitative
part with hard analysis, as shown in Table 1a.

Soft analysis is not so widely used in defence OA as
potentially hard analysis. One of the main methods of soft
analysis is the Analytic Hierarchy Process due to Saaty [77] ,
to which I shall return later. The main method of hard
analysis is what I refer to in the sub-title, and throughout
the paper, as the mathematical modelling of combat. The Saaty
method and the mathematical modelling of combat are shown as
examples of soft and hard analysis respectively in Table la.
My use of the term mathematical modelling of combat to describe
the main source of quantitative analytical experience is
explained below.

Analytical experience of combat dynamics may be obtained
from a variety of sources, such as: war itself, the analysis of
historical battles, the philosophical principles of war, the
theory of combat, combat modelling, war games, field trials
and military exercises. This list is an adaptation and
extension of that reproduced by Andrews and Laing [1] ,
following Low [63] , Taylor [86] and Hughes [45] . The
sources named vary widely in their realism, degree of
abstraction, reproducibility, convenience and cost. They are
complementary not competitive, a fact I emphasise by arranging
them in a loop as in Figure 2.

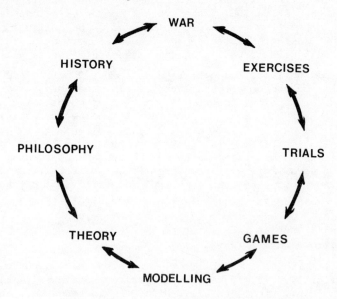

Fig. 2 Sources of Analytical Experience of Combat Dynamics

There are advantages in deriving analytical experience from more than one source. Many of the papers given at the International Symposium at Shrivenham in September 1984 covered several sources: for instance, Marshall [65] (history, philosophy and theory); Edwards [24] , Farrell [28] and Mertes [66] (theory and modelling); Callahan and Gagliano [10] (theory, modelling and games); Blumenthal [2] (modelling and games); Rowland [76] (modelling, games and trials); Koehler [52] referred to the whole range of sources from history to exercises. Adjacent sources may reflect strongly on each other and be especially powerful when combined.

Not all the sources, however, are strictly quantitative in the sense of providing a basis for hard analysis. In this paper I concentrate on two sources which are undoubtedly of a quantitative nature, namely combat theory and combat modelling. Without wishing to try to make precise definitions of the words theory and modelling, or run counter to the definition of theory given by Marshall [65] , in my usage I make three distinctions, as shown in Table 2. The first is that in a theory the essential analysis, the thinking, is done in symbols, while in modelling it is done in numbers. Secondly, theory is usually developed manually (through analytical models) while modelling requires the use of a computer. The third distinction is that theory involves solving a problem as part of the process of building the theory, whereas modelling requires production running which is an activity totally separate from the process of building a numerical computer model.

Given these distinctions, I take the mathematical modelling of combat to cover both combat theory and combat modelling, as shown in Table 2. It thus consists of those potentially quantitative sources of analytical experience of combat dynamics which do not involve human interactions. It does not necessarily require specific scenarios, or precise assumptions and data linked to particular weapon systems, or the need to form an analytical judgement to assist a planning decision. Its fundamental postulates may derive, in one direction, from war and the history and philosophy of combat or, in the other direction, from exercises, trials and games. The mathematical modelling of combat is a major tool for constructing a science of combat dynamics. Its division into two parts, shown in Table 2, is repeated in Table 1b.

The main form of <u>combat modelling</u> is by means of large, complex, mainly stochastic, simulation models. The attributes of good combat simulation models are comprehensively described in the book edited by Hughes [45] . The main form of <u>combat theory</u> is Lanchester theory, which is based on the solutions of sets of differential and difference equations. The book by Taylor [86] is the primary reference on Lanchester theory.

Table 2   Explanation of Term Mathematical Modelling of Combat

Possible Sources of Potentially Quantitative Analytical Experience of Combat								
War	History	Philosophy	Combat Theory	Combat Modelling	Games	Trials	Exercises
			Largely Symbolic	Essentially Numerical			
			+	+			
			Mainly Manual	Wholly Computerised			
			+	+			
			Model Solves Problem	Model Needs to be Run			
			Mathematical Modelling of Combat				
			Science of Combat Dynamics				

I should also mention here the model of land combat constructed
by Dupuy [23] , which consists of empirical formulae fitted to
historical data. It is a concentrated and illuminating source
of information on combat dynamics, perhaps more a calibrator
than a component of what I am calling combat theory. Its
relation to Lanchester theory is currently being explored.

Lanchester theory has been abundantly developed from the
original work in the famous book by Lanchester [55] . It has
not, however, become sufficiently coherent and comprehensive
to have attained the status of an accepted general theory of
combat. Curiously, its chief role in the mathematical modelling
of combat is not a direct one, as a theory, but an indirect one
as the basis of certain types of simulation modelling.
Nevertheless, it is in the direct sense that I include
Lanchester theory alongside simulation modelling in Table 1b.
Lanchester theory is described further in Section 5.

2.4 Organisation

OA in the Ministry of Defence (MOD) is done within a
structure which connects various groups in the Central Staffs,
the Armed Services and the Procurement Executive through a
series of committees. The committees sponsor, usually through
a nominated military officer of high rank, formal Studies
(which I emphasise by writing with a capital S). A Study is a
piece of analysis done in a limited time in answer to a
specific question and sponsored by the Armed Services to assist
the making of a decision. Studies are carried out by teams of
civilian analysts and military officers. The practice of
defence Studies provides the customer - contractor relationship
that is characteristic of OA, with the sponsor acting as a
surrogate for the decision-maker and the management of the Study
teams acting as the contractor.

The main centre for defence OA in the UK is DOAE, which
deals with the operations of all three Armed Services. DOAE
is dominated by the practice of defence Studies. It devotes
little effort to any analysis of defence operations which does
not have a clear link with current or projected Studies. For
later purposes I make a distinction between work done
essentially inside Studies and work done wholly outside Studies,
and set it out formally in Table 1c.

In a Study an attempt is made to bring the model world as
close as possible to the real world. Analytical judgement has
to be matched against the military judgement of the sponsor,
which stems from wide military experience. A Study therefore
involves the specification of detailed scenarios, orders of
battle, deployments of forces, and concepts of operations for

the weapon systems, sensors and support equipments concerned.
Such a specification requires many assumptions and their
associated data. A typical Study has hundreds of assumptions
and thousands of data items.

It is widely considered that these needs are best satisfied
by a methodology that is based almost exclusively on combat
simulation modelling. DOAE, for instance, has a long-term
programme of building large, complex, computerised, mainly
stochastic, combat simulation models for use in Studies. The
domination of the methodological process by a single analytical
method then tends to dictate the form of the organisational
procedure through which a Study is conducted. The procedure
for conducting Studies at DOAE, described in a memorandum by
the Director [22], illustrates this point. It identifies
three phases in a Study.

The first phase is for setting up and defining the Study;
this involves accepting a request for a Study on a topical
question, formulating the Study question in analytical terms,
selecting (or if necessary building or having built) an
appropriate simulation model, defining the specific Study
assumptions and collecting the input data. The second phase is
the phase in which the bulk of the work is done and the Study
report is written; this involves feeding the assumptions and
data into the model, running the model, analysing the output,
presenting the results to the sponsor verbally, and then
writing the Study report. The third phase is for agreeing the
Study report with the sponsor and then publishing it.
Publication of the Study report marks the end of a Study. DOAE
is not involved in the making or implementing of any subsequent
decision. To that extent defence OA does not cover the full
process of traditional OA.

Studies are generally regarded as being limited in time. In
fact, the first and third phases are of unspecified duration;
only the second, the doing and reporting, phase has a fixed
time limit. However, the feeling of working against time is
invariably present. Only rarely does the time ostensibly
available appear to be remotely commensurate with the magnitude
of the job to be done. Studies are conducted in a hostile
environment; trying to cope with big issues and bewildering
details with one eye on the clock leaves little opportunity for
rumination and reflection.

2.5 *Some Major Points*

By making simplifications in the above survey, I have
concentrated attention in Tables la, lb and lc on three major
distinctions connected in turn with the definition of defence

OA, the mathematical modelling of combat, and the way defence
OA is organised. These distinctions are superimposed in Table
1d to identify four principal components of analysis. Since
soft analysis is designed for use inside Studies, and combat
modelling is not likely to be used outside Studies, the four
resultant components are: soft analysis (particularly the Saaty
method) inside Studies, combat modelling (particularly
simulation modelling) inside Studies, combat theory (particularly
Lanchester theory) inside Studies, and combat theory (particularly
(Lanchester theory) outside Studies.

However, the survey has stated that: (i) defence OA is
dominated by Studies; (ii) Studies are dominated by combat
simulation modelling; (iii) Lanchester theory is not an
accepted general theory of combat; (iv)' where Lanchester theory
is applicable it is only rarely used directly in Studies; (v)
soft analysis is little used in Studies. Hence, of the principal
components of analysis identified above, at present the
overwhelming effort goes on combat modelling inside Studies, and
particularly on simulation modelling. But combat simulation
modelling in Studies, the predominant form of the mathematical
modelling of combat in defence OA, encounters formidable
problems and has been subjected to severe criticisms. The next
section indicates what some of the problems and criticisms are.

3. SOME CRITICAL COMMENTS

3.1 Problems and Criticisms

In discussing problems and criticisms of combat simulation
modelling in Studies I find it convenient to start from the
article by Wood [100]. Wood cites earlier critical papers
from the USA by Bonder [4], Shubik and Brewer [80] Stockfisch
[82], Low [63], and the Comptroller General of the United
States [16]. To this list I add the paper by Strauch [84].
These American papers contain references to many others, and
also a great deal of constructive material. They give an
extremely detailed critique of combat simulation modelling and
of quantitative methodology in general which applies to the
present situation in the UK.

Wood [100] expresses five particular grounds for unease.
They are: (i) the lack of an accepted theory of combat; (ii)
the ambiguous part played by judgement in models; (iii) the
lack of good data; (iv) the problem of dealing adequately with
the true level of uncertainty in assumptions and processes;
(v) the extreme difficulty that the decision-maker has in
assessing the real significance of Studies based on models.

The problem of uncertainty was one of three fundamental
problems discussed at a seminar in 1978 conducted by the Chief
Scientific Adviser MOD [12]. The other two problems were
those associated with the wide scale of operations in combat
dynamics and with the elusive nature of suitable criteria
(measures of effectiveness, for instance) for expressing the
results of combat simulation modelling. I wish to include
these two problems also, and since they are more closely linked
to the problem of uncertainty than to the other four of Wood's
grounds for unease, I shall expand the problem of uncertainty
to become the problems of uncertainty, scale and criteria.

Wood's grounds for unease are consistent with the implied
criticisms in my survey of defence OA. By considering them
together, amalgamating some and re-interpreting others, and
rearranging their order, I arrive at a list of problems and
criticisms of the use not just of combat simulation modelling
in Studies but of combat modelling and combat theory, that is
of the mathematical modelling of combat, in defence OA. The
list is as follows.

(1)   The mathematical modelling of combat, although not in
      itself a method for helping to make decisions, is not
      usually accompanied by any soft analysis in Studies.

(2)   Combat modelling is used to excess in Studies, and its
      quality may not always be sufficiently high because:

      (a)   combat simulation models are often too large, complex
            and opaque;
      (b)   there is a lack of good data, with the emphasis on
            the word good;
      (c)   there are no agreed ways of treating the problems of
            uncertainty, scale and criteria.

(3)   Although combat modelling is a quite different thing from
      a theory of combat, it is not usually supported in Studies
      by the available combat theory.

(4)   The theoretical material that exists on combat dynamics is
      not being developed outside Studies into a coherent and
      comprehensive theory of combat.

3.2 Observations

The practice of defence Studies contains a remarkable
paradox. Sponsors of Studies often insist on the most minutely
detailed inputs to the analysis for the sake of operational
credibility, while only being prepared to accept the grossest
interpretation of the outputs because operational flexibility

is an overriding military principle. They do so also because
they believe that without detailed inputs even gross
interpretations may be erroneous. But imagine a customer,
requiring cream, insisting that a farmer should collect a ton
of grass, submit each blade for inspection, and then feed it
all into a cow, instead of getting the cream from the dairy.
The sponsor, however, has an excuse: defence OA does not have a
dairy, because there is no established science of defence
operations, not even of combat dynamics.

The lack of an accepted theory of combat, in particular, is
at the roots of the difficulties of a Study team. It has been
highlighted by the recent publication of a paper by Weiss [93] ,
written thirty years ago and still relevant. Studies are not
built on a solid quantitative foundation of past work but are
individual pieces of analysis starting more or less from
scratch. The task of the Study team is often to do a piece of
analysis around one or two base-case scenarios in a manner
which is ad hoc with regard to the Study question and expedient
with respect to time.

The Study team are, in effect, required to undertake some
numerical experiments, good things in themselves, almost as if
they were doing research in a physical science. But, since in
defence OA there is no generally-accepted background theory
and usually no advantage is taken of what theory there is,
the experimental points stand alone. Not even the simplest
forms of theory are used. There does not exist, for instance,
a standard notation or a basic set of non-dimensional
parameters, used by everybody everyday. Moreover, the results
of the numerical experiments are only rarely consolidated or
collected together for future use. Just compare defence OA
with physics itself. Where would physics be if every physicist
had to discover the laws of physics for himself? It is still
necessary to heed the exhortation of Koopman [53] to improve
'the basic education of operational research workers in the
methods, points of view, and habits of thought, which have for
so long been current in mathematical physics'.

The point at issue is the balance that should be struck
between the search for operational credibility and the
achievement of analytical understanding. I submit that in the
UK the emphasis is placed so much on trying to produce
detailed operational credibility that in many Studies
analytical understanding is effectively lost.

3.3 Analytical Experience and Analytical Judgement

What does this situation amount to in terms of the analytical
experience gained in a Study and the analytical judgement formed
from it? Consider the distinction between quantitative and
qualitative analytical experience. Qualitative analytical
experience is gained, for example, by building combat
simulation models. By contrast, quantitative analytical
experience, which requires the assimilation of quantitative
answers to operational questions, comes from the results of
running simulation models and, especially, from theories.
Clearly, quantitative analytical experience cannot be acquired
at all until an adequate level of qualitative analytical
experience has been attained.

Quantitative analytical experience is a rather rare
commodity, for two reasons. First, the quantitative analytical
experience of a given Study question that may be accumulated
prior to a Study is usually very limited because of the
absence of an accepted theory of combat and because, in the UK,
combat simulation models are scarcely ever run outside Studies.
Secondly, the quantitative analytical experience gained during
a Study, being derived from a few numerical experiments by
combat simulation modelling, is also likely to be very sparse.
Moreover, if the simulation model is not wholly transparent,
the results obtained by running it may be of doubtful validity.
Quantitative analytical experience in Studies may therefore
be both scant in quantity and dubious in quality.

Consequently the analytical judgement that is the outcome
of a Study is likely to be based on a narrow and possibly shaky
quantitative foundation. In fact, I contend that it is
qualitative analytical judgement that holds sway in defence
Studies, in spite of the quantitative analytical techniques
that are used in arriving at it. But if defence OA is to be
regarded as a quantitative subject it should be quantitative
analytical judgement, based on wide quantitative analytical
experience, that is offered.

It is sometimes said in mitigation of the current practice
of defence Studies that it is the actual doing of a Study, rather
than its outcome, that provides the benefit to the Armed
Services. There is some truth in the argument that a Study is
a vehicle for a useful debate that might not otherwise take
place. But a Study cannot be of any lasting value if its
results are not worth preserving and promulgating. And surely
the argument does not justify the excessive lengths to which
combat simulation modelling is often taken. There is force in
the counterargument, contained in the critical papers cited
earlier, that defence OA should either do a better job of
providing quantitative analytical judgement or relinquish the

historical concept that quantitative analysis is part of its
function.

3.4 Assessment

Defence operational analysts have an extraordinarily
difficult task, almost an impossible one. Too much should not
be expected therefore. But, being based on combat simulation
modelling within time-limited Studies on specific questions,
defence OA can be expensive in effort and facilities without
providing reliable quantitative analytical judgements to assist
the making of planning decisions. Too often, through sins of
omission as much as commission, defence OA is neither good
operational analysis nor good science, when it should be both.

To achieve both operational credibility and analytical
understanding in defence Studies, the mathematical modelling of
combat needs to be put in a more favourable context (better OA)
and to be made to produce more authentic results (better
science). There seems to be a case for an approach to Studies
which differs from the current approach in three respects: in
its attitude of mind, in the mixture of analytical methods it
employs, and in the procedure by which the methods are put into
practice. Some general suggestions about each of these aspects
are made in the next section.

4. SOME CONSTRUCTIVE SUGGESTIONS

4.1 Attitude of Mind

The key to a different attitude of mind towards Studies
seems to me to lie in a combination of two crucial features of
defence OA. The first is the apparently paradoxical requirement
of the sponsor of a Study for detailed inputs and gross
outputs. The second is the inevitably uncertain nature of the
input data. In such a situation I suggest that the best that
can be hoped for as the outcome of a Study is a balanced
quantitative analytical judgement that has the properties
described below.

First, the relation of analytical judgement to military
judgement should be made clear. Analytical judgement is
complementary to military judgement and is to be compared with
it. Analytical judgement must be presented with absolute
clarity and conviction if it is to influence military judgement,
which (leaving aside political judgement) is the final arbiter
of an analytical judgement. If an analytical judgement cannot
be defended convincingly against the, possibly preconceived,
military judgement of the sponsor then the analytical judgement
should surely be accepted as subordinate to the military

judgement. For all the prejudices that may be involved in
the military judgement, it may well be based on more extensive
experience than the analytical judgement, and the military have
to live with any decisions that follow from it. Analytical
judgement assumes its greatest importance when there is no
consensus of military judgement, say between different Armed
Services.

Second, quantitative analytical judgement should be derived
from a blended quantitative analytical experience that is
itself tailored to the nature of the judgement to be drawn from
it. The primary factor here is that the quantitative analysis
which provides the analytical experience in a Study should seek
as outputs only gross statements to single figure accuracy.
(This suggestion is akin to the famous 'hemibel thinking' of
Morse and Kimball [67], but on a rather finer scale). Analysis
should be concentrated on the border-line between quantitative
and qualitative, where figures may be easily translated into
words, experience into judgement. While the quantitative
analysis should include some special cases, based on detailed
inputs, to demonstrate that the gross output statements hold
good in specific circumstances of outstanding operational
interest, the role of the special cases should be to provide
confirmation rather than inspiration. The operational
credibility of a Study should be measured more by the success
of its analytical judgement in stimulating the sponsor than by
the copiousness of the details of its analytical experience.

Third, to increase analytical understanding, a form of
control should be exercised over the total analysis done in a
Study. This could be achieved by employing a mixture of
analytical components within a procedure which provides a hard
scientific core and a soft judgemental cover. The part played
by judgement would then be more clearly defined and the
demands on the quantitative analysis made less severe. The
next two sub-sections put forward some suggestions along these
lines.

4.2 Mixture of Methods

Table 1d identifies four principal components of analysis.
To repeat, they are in general terms: soft analysis inside
Studies, combat modelling inside Studies, combat theory inside
Studies, and combat theory outside Studies. Now consider these
four components as the basis for a mixture of methods. It is
already clear that there is a good case for soft analysis
inside Studies to cover the decision-making part of defence OA.
To supplement combat modelling by a judgemental technique
would be a straightforward way of helping the sponsor to
appreciate the significance of the modelling results.

There is also a strong case for combat theory inside Studies. It runs as follows: there is a lack of good data, therefore a parametric treatment of data is required, hence wide sensitivity will be revealed, therefore symbols (which may automatically and instantly display sensitivity) would help, therefore theory (which is founded on symbols) should be used. But, since theory cannot be done easily or thoroughly under the constraints of a Study, there is also a case for combat theory outside Studies.

Indeed, as already implied, the main development of combat theory should take place outside Studies and the results should be applied later in Studies. The fundamental difficulties of uncertainty, scale and criteria can only be tackled properly outside Studies. But, in the UK, the lessons of combat theory will only be put to use through an analytical judgement matched against a military judgement in a Study. Ideally, therefore, combat theory should be developed outside Studies in such a way that its results are directly applicable inside Studies.

Hence it is possible to imagine a hypothetical alternative to the present situation, as typified by that at DOAE, in which practically all analytical effort goes into combat simulation modelling inside Studies. The hypothetical extreme would have, say, half the effort devoted to building up combat theory outside Studies and the other half to doing soft analysis inside Studies. In this imagined extreme, defence operational analysts would possess so much quantitative analytical experience from the information available outside Studies that they need only exercise their analytical judgement, with the aid of some soft technique, when called upon to do specific Studies.

Although the hypothetical alternative might be an improvement on the present situation, neither of the two extremes provides a satisfactory distribution of effort. So consider how the total analytical effort might be divided between all four of the principal components of analysis identified in Table 1d. Let the effort be plotted on a graph which ranges from the present extreme on the left to the hypothetical extreme on the right. Consider the division of effort for each major distinction in turn.

In Figure 3a, curve I is a dividing line between soft analysis and hard analysis. In Figure 3b, curve II divides the hard analysis into combat modelling and combat theory. In Figure 3c, curve III divides the total effort into that expended inside Studies and that expended outside Studies. The curves are parabolas, which are the simplest curves

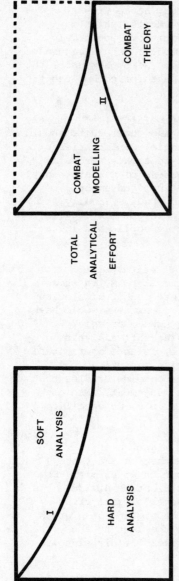

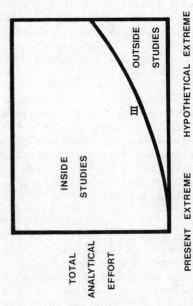

Fig. 3a First Major Distinction

Fig. 3b Second Major Distinction

Fig. 3c Third Major Distinction

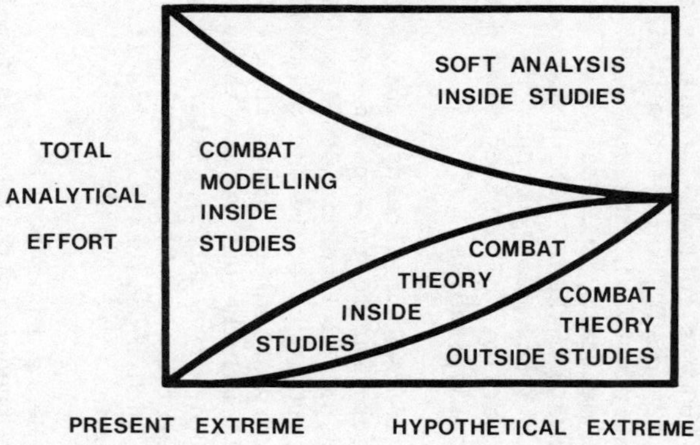

Fig. 3d Four Principal Component of Analysis

with suitable properties near to the extremes. Figure 3d is
the result of superimposing Figures 3a, 3b and 3c. When
followed from left to right the curves show soft analysis
inside Studies and combat theory outside Studies gradually
increasing, with combat modelling inside Studies gradually
decreasing to zero and combat theory inside Studies rising from
zero to a maximum in the middle and then decreasing to zero
again.

I suggest that a better division of analytical effort than
the present would be got by moving substantially to the right
towards the hypothetical extreme. For example, taking the
position midway between the extremes in Figure 3d gives the
division of effort between the four components (soft analysis,
combat modelling and combat theory inside Studies, and combat
theory outside Studies) in the ratios 3:2:2:1. This is not
an unreasonable starting point for discussion. Based on it,
my main suggestions are therefore for:

(1)   more use of soft analysis inside Studies;
(2)   less use of combat modelling inside Studies;
(3)   more use of combat theory inside Studies;
(4)   more development of combat theory outside Studies.

As far as magnitude of effort is concerned, suggestions (1)
to (4) above correspond with the criticisms (1) to (4) given
earlier. Quality of work is a further issue.

Taken together, combat modelling and combat theory may be
combined into an empirical form of mathematical modelling
inside Studies. The question then arises as to how such an
empirical method might be linked with a soft judgemental
technique to create an acceptable organisational procedure within
a Study. This question is addressed in the next sub-section.

4.3 *Procedure for Doing Studies*

When, as at present, Studies are dominated by the single
method of combat modelling, the organisational procedure
follows as a matter of course. However, when a mixture of
methods is to be employed, the time sequencing of the methods
becomes important. While it would not be sensible to have a
procedure that is too cut and dried, there are several
considerations that tend to prescribe a certain order to the
required stages. One is to separate the gaining of analytical
experience (the hard core) from the exploring and forming of
analytical judgement (the soft cover). Another is to contain
the combat modelling within a framework set by combat theory.
Ideally, the latter would itself be a slice of a general theory
of combat dynamics developed outside Studies.

I therefore suggest that the first two phases of a Study
(for setting up and defining, and doing and reporting) might be
joined together and then divided into five stages, prior to
the drafting of the Study report, as follows:

(1)   an initial judgemental stage for getting to grips with
      the Study question and seeking a provisional judgement;

(2)   an initial theoretical stage for finding out what is
      already known and for planning the runs of the combat
      simulation model;

(3)   the modelling stage;

(4)   a second theoretical stage for consolidating and
      generalising the results of the simulation modelling;

(5)   a second judgemental stage for blending the various
      strands of analytical experience and forming the balanced
      analytical judgement.

At times different members of the Study team might well be
working in different stages.

Table 3 gives a compact view of the features of the above
procedure (and also anticipates some suggestions to be made
later). From the top downwards it shows the breakdown into
stages of the total analysis, without trying to set the
individual amounts in particular proportions as might be
derived from Figure 3d. The hard core aims to provide a blended
quantitative analytical experience by means of controlled,
rather than ad hoc and expedient, mathematical modelling of
combat. The actual method of control consists of a combination
of symbolic analysis and numerical experiments.

The case for a controlled empirical form of the mathematical
modelling of combat is very strong. Neither symbolic combat
theory nor numerical experimentation by combat modelling is
good enough alone. But each is best at what the other is
worst. An empirical method could make the most of both.
Combat theory could provide the big picture, the trends and
sensitivities, perhaps even the clue to the analytical
judgement that would stimulate the sponsor. Combat modelling
could be used to satisfy the input requirements of the sponsor
and yield the detailed evidence that would help the judgement
to carry conviction with him.

From the bottom upwards, Table 3 indicates how the
transitions between stages could mark the changes that take
place in the character of the work done during a Study, where

Table 3 Features of Suggested Procedure for Doing Studies

Total Analysis in a Study				
Soft Cover	Hard Core			Soft Cover
Balanced Quantitative Analytical Judgement	Blended Quantitative Analytical Experience			Balanced Quantitative Analytical Judgement
	Controlled Mathematical Modelling of Combat			
	Empirically Enhanced Symbolic Analysis			
	Symbolic Analysis	Numerical Experiments	Symbolic Analysis	
Initial Judgemental Stage	Initial Theoretical Stage	Modelling Stage	Second Theoretical Stage	Second Judgemental Stage
Soft Analysis	Combat Theory	Combat Modelling	Combat Theory	Soft Analysis
Saaty Method	Lanchester Theory	Simulation Modelling	Lanchester Theory	Saaty Method
Pairwise Comparisons	Non-dimensional Parameters	Parity Condition	Non-dimensional Parameters	Pairwise Comparisons

← Empiricism →

Transitions between Experience and Judgement

Transitions between Analytical Judgement and Military Judgement

Transitions Requiring Recognisable Compromises

recognisable compromises are needed. The transitions divide
the impreciseness (or messiness, fuzziness, squishiness as it
is variously called) of a Study into three separate
compartments. First, at the beginning and end of a Study there
are transitions between military judgement and analytical
judgement. Second, early in the Study and again later there
are transitions between analytical judgement and analytical
experience. Third, on occasions during the period devoted to
quantitative analysis there are transitions between theory and
modelling which contain the elements of empiricism. These
transitions could help the Study team to recall the exemplary
instruction given in 1979 by the then Director of DOAE,
Dr. I.J. Shaw [22] : '.. do not allow the beauty of the
mathematics, or the subtlety of the model, to blind you to the
fact that you are dealing with potentially real hardware, real
tactics, and real operations ..'.

Distributing the analytical effort among the different
methods, as indicated in Figure 3d and Table 3, should benefit
the overall quality of the analysis done in Studies. But
changing the relative amounts of work done in the various
components of analysis does not necessarily guarantee a high
standard of work in each component. This topic is the concern
of the second part of the paper (Sections 5,6,7), which gives
some more detailed measures related specifically to the Saaty
method, Lanchester theory and simulation modelling. The
measures are developed from some features of Lanchester theory
which are described first to provide the necessary background.

5. SOME FEATURES OF LANCHESTER THEORY

5.1 Description

I take Lanchester theory to cover the analysis of combat
through the solutions of sets of differential and difference
equations by means which include the sustained manipulation
of symbols. As described, the term Lanchester theory applies
to most of the works cited by Taylor [86] , and to most of the
references compiled by Daly [18] for the informal UK Lanchester
Study Group (see Daly [19]). Daly's list, an extension of
the original compilation by Haysman and Mortagy [37] , has
grown to over six hundred titles. The treatise by Taylor [86]
brings together a wealth of analytical results previously
scattered through hundreds of separate papers, and it is now
the natural starting point for any appraisal of Lanchester
theory. Also of great interest, but less accessible, are the
collected papers by Willis [96,97,98] .

Lanchester theory is closely akin to system dynamics (see
for instance Forrester [29] , Coyle [17] and Wolstenholme

[99]), but with more emphasis on the symbolic nature of the analysis. The Lanchester differential equations are also similar to those of chemical kinetics and to some occurring in the mathematics of biological systems. They are not a closed set as the Navier-Stokes equations in fluid dynamics are. I do not know whether they have ever been set out in their greatest possible generality. Naturally, Taylor [86] goes a long way in this direction, and the computer model recently described by Sassenfeld [78] is based on a generalised form of Lanchester equations. But from the symbolic point of view Lanchester theory is largely a collection of classic solutions to simplified problems.

My description of Lanchester theory stresses its symbolic basis. Symbolic analysis has distinctive advantages over the use of purely numerical analytical methods. By the algebraic manipulation of symbols and the introduction of devices of approximation, a piece of analysis may be worked through to a solution which may then have the merits of permanence and commonality between sea, land and air operations. The solution may display the sensitivities of the problem without the need for numerical calculations. Aggregated measures and criteria are a natural byproduct of symbolic analysis. So are non-dimensional parameters. These features are not normally present in analysis performed numerically.

While I regard Lanchester theory as a form of mainly symbolic mathematics, the difference-differential equations in stochastic analyses of combat theory (as distinct from combat modelling) are too complex to be solved analytically except in very special cases (Bowen [6,8] , Willis [98] and Hynd [46] give interesting examples). In general, numerical methods must be used to complete the process of solution. Weale and his associates [89-92] have constructed several meticulous numerical solutions of this type, but the solutions are not applied as widely as they deserve to be. Karr [50] , following many penetrating papers on the rigorous basis of stochastic Lanchester theory, has examined the use of stochastic Lanchester equations in the construction of large computerised combat simulation models in the USA. In the UK, the use of equations of Lanchester type as building blocks in computer simulations has been developed by Haysman and his colleagues in a way which combines many of the most powerful features of symbolic Lanchester theory; see for instance Mortagy [68] , Haysman and Wand [39] and Read, Wand and Haysman [74] .

Lanchester theory thus includes wholly symbolic and partly numerical treatments, and covers both stochastic and deterministic analyses of combat. It involves discrete and continuous variables, rigorous and heuristic assumptions, exact solutions and particular approximations. It applies to both homogeneous and heterogeneous forces. The Lanchester equations may be written with three space coordinates as independent variables, or with spatially - dependent coefficients. But one of the great strengths of Lanchester theory is that it can provide aggregated results from which the space coordinates have been eliminated or, for instance in land combat, represented by the movement of the front line. The principal variables are usually time and the force levels of the combatants. The latter are often referred to as the variables of state.

Lanchester's original work [55] made use of deterministic analysis with continuous variables, through differential equations. It contained his famous square and linear laws, for the direct-fire and indirect-fire battles respectively. My description of Lanchester theory covers also the solution of the deterministic difference equations which are the symbolic generalisation, given by Engel [26], of the model of combat due to Fiske (as described by Weiss [94]). The continuous Lanchester square law follows from the solution of the discrete (in time) Fiske model in the limit of small attrition rates. This result supports the related statement made by Neild [70]. The effects of the differences between using continuous and discrete variables can be quite marked, and they are sometimes not distinguished from those due to the differences between deterministic and stochastic treatments. There are still some fundamental issues and many mathematical niceties to be resolved in Lanchester theory.

5.2 Combat Effectiveness

There are expressions which are, in the deterministic case, and whose expected values are, in the stochastic case, invariant with respect to time during combat. The combat invariants are part of the solution, say, of the course of a battle. They give rise to measures of combat effectiveness. They include a definable measure of the relative combat power of two forces which takes account of both the firepower and the tactics of the combatants (see Hynd [46]).

When the initial conditions at the start of a battle are applied to the combat power invariant the resulting equation is called the equation of state of the forces of the two sides. The equation of state may be written in several algebraic forms. When written as a single expression equal to zero the expression may be called the combat power function.

For particular strategies and tactics, the combat power function separates into the difference between the combat powers of the two sides. This occurs, for instance, in the direct-fire battle treated by continuous deterministic differential equations. In this case it leads to the usual form of the Lanchester square law as given by Haysman and Wand [39]. In general, the deterministic equation of state, in whatever form, allows the force level of one side to be calculated when that of the other is assumed.

However, it is possible to give a different twist to the equation of state by introducing pre-conditions for the end of combat: the termination or breakpoint conditions (see Taylor [86]). The termination conditions may be thought of as representations of the will to fight of the two combatants. In the deterministic case, when the termination conditions are expressed in the form of pre-set defeat levels and are applied to the equation of state the result is the condition for a tie, otherwise known as the break-even point or parity condition. The parity condition gives an output relationship between all the input parameters; it thus determines the parity value of one of them when the others are specified. Alternatively, it may be interpreted as giving the parity value of the force ratio, for instance. Note that the often-made assumption that each side is prepared to fight to annihilation gives only one special (and highly unlikely) case of the parity condition. When arbitrarily chosen numerical values for the initial and termination conditions are applied to the combat power function its value will not in general be zero. A non-zero value of the combat power function indicates that one side or the other has the advantage, and hence the function may be referred to as the combat power advantage.

When termination conditions are specified as part of the input information in order to complete the specification of a battle, it is possible in stochastic battles to define and determine the combat power advantage and the probability of victory as measures of combat effectiveness. (In deterministic treatments, the probability of victory is either zero or unity, changing from one to the other at deterministic parity, that is at the tie point.) The concept of the probability of victory is a very attractive one, but the actual calculation of it in terms of the input parameters is extremely difficult in all but the simplest cases. Even where formulae do exist they do not always impart numerical insight at first glance.

5.3 Parity Condition

The parity condition determines the circumstances in which two combatants are evenly matched. In effect it reduces the number of input parameters by one, but its great merit lies in

its providing the dividing line between victory and defeat.
It thus sets aside a double infinity of less clear-cut cases.
The parity condition is particularly useful for the purposes
of planning, and is one of the central results of Lanchester
theory from the point of view of a Study team.

The most obvious form of parity condition is that which
gives an equal probability of victory to each side; that is a
probability of a half in cases where a tie would occur in the
deterministic analysis. I shall refer to this form of parity
as probability parity, and use the name power parity for the
form of parity condition obtained when the combat power
advantage, that is the combat power function in terms of all
the input parameters, is put equal to zero. In principle,
deterministic parity, power parity and probability parity are
all different.

A numerical comparison of deterministic parity, power
parity and probability parity for the case of the Lanchester
square law battle is given by Lord [61] . This illustration
was stimulated by the work of James [47] and Hartley, Hagues
and Kettle-White [35] . It made use of the work of Hynd [46] ,
Watson [88] , Weale [89] and Gye and Lewis [34] . It shows
that, over the full ranges of the governing non-dimensional
parameters, there is little difference between probability
parity and power parity and that both are sufficiently close
to deterministic parity for the latter to be seen as an
adequate first approximation.

Economically, in terms of the number of concepts to be
considered in practical applications of Lanchester theory, there
might be a case for concentrating on power parity at the expense
of probability parity in stochastic treatments. While
probability parity is perhaps easier to comprehend, it is not
connected so smoothly analytically with deterministic parity.
But the probability of victory is such an attractive concept,
especially to military officers, that I shall continue to give
it equal status with the combat power advantage as a measure
of combat effectiveness.

5.4 Non-dimensional Parameters

I have noted above that the comparison of the different
forms of parity made use of non-dimensional parameters. Non-
dimensional parameters are naturally-occurring dimensionless
combinations of several individual dimensional parameters.
They do not in themselves solve anything, but they economise,
concentrate and simplify, and thus they help to create
analytical understanding. They may be otherwise known as
similarity parameters, portmanteau parameters, unit-less

parameters, or scale factors. Under the last name they have
been extolled by R.V. Jones [48] : 'I have been so struck by
scale factors in various fields that I think that it is worth
running the danger of over-labouring the point.'

To introduce some particular non-dimensional parameters,
consider a potential battle between two homogeneous sides Red
and Blue. Assume that a Red force is thought to be preparing
to attack some Blue targets. Red thus poses a threat to Blue
and Blue therefore devises a deterrent to, or a defence
against, Red in the form of a Blue force to counter some Red
targets. The Red targets for the Blue force might be the
threatening, or attacking, Red force itself (as in the direct-
fire battle) or some totally different targets. Similarly, the
deterring, or defending, Blue force could be the original Blue
targets (again, as in the direct-fire battle) or some entirely
new force. Whatever the situation, the analysis of the
potential battle will involve the specification of the
individual effectiveness of a component of the Red force
against a Blue target and the individual effectiveness of a
component of the Blue force against a Red target. Once the
individual effectiveness of a force has been expressed in
suitable units of targets per force, the product of the
number of the components of the force and the effectiveness of
a single component divided by the number of the targets yields
a non-dimensional parameter.

In this way, the Red force and the Blue targets give rise to
a non-dimensional Threat, or Attack, parameter, and the Blue
force and the Red targets lead to a nondimensional Deterrent, or
Defence, parameter. These non-dimensional parameters are formed
from dimensional input parameters related to the initial
conditions. Their ratio is similarly a non-dimensional parameter
related to the initial conditions. It is an important ratio
which recurs throughout the theory of combat dynamics. For
definiteness in what follows I shall henceforth consider the ratio
of the Defence (or Deterrent) parameter to the Attack (or Threat)
parameter, that is: Defence parameter/Attack parameter.

In the case of the Lanchester square law (direct-fire)
battle, Helmbold, in past work cited by Taylor [86] , has called
the logarithm of the square root of this ratio the 'defender's
advantage parameter'. He has recently made striking use of this
parameter, in an empirical analysis (Helmbold [40]) of a large
number of land battles drawn from the historical data base of
Dupuy [23] . Using logistic regression to determine, among
other things, the probability of a defender's victory from the
information on the defender's advantage parameter and the known
battle outcomes, Helmbold finds that the empirical advantage
parameter seems to capture most of the factors associated with

victory.   Thus an encouraging start has been made to a further
approach to the analysis of historical battles.

Taylor [86] himself calls the square root of the ratio of
the Defence parameter to the Attack parameter in the case of
the Lanchester square law battle the 'normalised initial force
ratio'.   The names given by Helmbold and Taylor help to
identify the nature of their parameters, but perhaps fundamental
non-dimensional parameters in defence OA would benefit from
having short, easily remembered, personalised names, as in
fluid dynamics.   I have in the past (Lord [60]) called the
inverse of the Defence parameter/Attack parameter the Lanchester
number.   Helmbold's logarithm, his and Taylor's square root and
my inverse, while not of fundamental importance, serve to
demonstrate the need for a common notation and terminology.
I shall therefore abandon my inverse and, taking the point of
view of the defender, in this paper call the ratio Defence
parameter/Attack parameter the Lanchester number.

As observed earlier, to define a potential battle fully it
is necessary to specify some factors, the termination
conditions, which determine the end of combat.   Given the
solution for the course of the battle, the parity condition
connects these termination conditions with the initial
conditions.   The deterministic parity condition can, in
principle, be expressed in terms of the parity value of the
Lanchester number, which will be a function of the other
non-dimensional parameters of the problem.   When the parity
value of the Lanchester number is known it is possible to
define a further non-dimensional parameter by dividing the
Lanchester number by its parity value.

The resultant non-dimensional parameter is a measure of
the 'numerical factor that expresses the comparative value of
two contending forces', sought so long ago by Admiral Fiske.
I therefore suggest that this parameter might suitably be
called the Fiske number.   It could be denoted, following the
tradition of fluid dynamics, by the symbol Fi.   To give an
example, for the continuous deterministic Lanchester square
law battle with dimensional parameters B and R (initial force
levels), β and ρ (effectiveness parameters), and b and r
(defeat levels), the Fiske number is given by the formula

$$Fi = \beta(B^2 - b^2)/\rho(R^2 - r^2).$$

The parity value of the Fiske number, Fi*, say, is unity when
evaluated by the method by which the solution of the battle
is derived.   It is likely to be close to unity when evaluated
on some other related basis, for instance by stochastic power
parity or probability parity.   (This point is illustrated in

Lord [61], in which paper I had not yet arrived at the name
Fiske number or the symbol Fi). Once a Fiske number is
determined in its functional form, input values of the
dimensional parameters which lead to a numerical value of the
Fiske number greater than unity indicate circumstances in
which the defender would have the advantage. The Fiske number
is an ideal parameter against which to plot some other output
non-dimensional parameter, such as the probability of victory,
as described in the next section.

6. PARTICULAR MEASURES FOR USE INSIDE STUDIES

6.1 Soft Analysis (Saaty Method)

6.1.1 General Preliminaries

The objective of the initial judgemental stage of a Study is
to get a quantitative perspective view of the Study question,
perhaps even an inkling of the answer. The first requirement
is to identify the principal level of operations, which must
then be associated with the higher level to which it
contributes and with the lower level operations which contribute
to it. Identifying such levels is not easy. The hierarchy of
operations is more like a spiral than a neat tier. It is
particularly difficult to construct a hierarchy with components
at all levels which are analytically understandable, separately
and together.

The second requirement is to regard the analysis
parametrically, being prepared to use plausible ranges of data
rather than just precise values. By adopting a parametric
treatment of data there should not be any need to feed numerical
information directly from one level of the hierarchy to
another. Indeed, the aim should be to dispense with the
hierarchy once the ranges of the data inputs to the principal
level of operations, and the form of the required outputs, have
been established.

Even with a parametric treatment of data, a Study will still
need to be set in one or more detailed scenarios, with precise
assumptions and data specified to give an air of realism to the
simulation modelling. Such intermediate situations provide
spot points which need to be seen against an appropriate
background. While plausible ranges of data help to provide
local consistency and sensitivity, a more definite framework
for the analysis is required. It is therefore helpful to
identify and examine some significant extreme cases which are
tractable analytically. These serve to indicate the bounds of
the analytical possibilities.

6.1.2 Pairwise Comparisons

There are many soft analytical techniques that might be used in the judgemental stages of a Study. From my personal experience the method of Saaty [77] has several attractive properties. It has been used with success in military problems for some time by Haysman [38] , and its application in defence OA has increased markedly in recent years. At the Shrivenham Symposium of 1985, for instance, papers connected with the Saaty method were presented by Hartley and Haysman [36] , Hough and Isbrandt [42] , Godfrey and Anderson [30] , and Everett, Harmsworth, Miller and Hanbury [27] . There was general agreement that the method, though not without its defects and difficulties, is of great assistance in gaining insights into Study questions.

An important step in the Saaty method is to interpret the Study question as a decision between alternative choices with specified attributes. The number of choices at the principal level of operations should be reduced to the absolute minimum. So too should the number of attributes by which the choices are to be judged. This concentration of attention is of the utmost importance if any semblance of analytical understanding is to be achieved in a time-limited Study. Many Study questions, when first posed, are far too wide in their scope and, at the same time, too detailed in their implications. It may not be possible to arrive at the most appropriate selection of choices and attributes in the initial judgemental stage, but it is essential to start off with as few as possible. There will be opportunity for adjustment during the course of the Study, perhaps even in the second judgemental stage.

The Saaty method should be applied in its simplest form at the principal level of operations. The Choice Attribute Ranking Package of Lockett (Lockett, Muhlemann and Gear [58] describe an application) seems ideal for this purpose. From the point of view of the present paper the Saaty method has two especially advantageous features: it employs pairwise comparisons (of attributes and of choices with respect to each attribute) , and it uses a comparison scale which effectively involves only single figures. It provides a bridge with qualitative methods, such as that described by Cockle [14] at the 1985 Shrivenham Symposium, and fits in perfectly with the attitude of mind towards Studies put forward in this paper. In Table 3, I single out the technique of pairwise comparisons as one of the major measures for use in Studies. A list of all the main suggestions made in this sub-section and in the rest of the paper is given in Table 4.

Table 4   List of Main Suggestions

Approach to Studies

(1)   regard military judgement as the arbiter of analytical
       judgement;
(2)   seek statements of analytical experience to single figure
       accuracy only;
(3)   separate the soft cover of analytical judgement from a
       hard core of analytical experience;
(4)   derive analytical experience from an empirical form of
       mathematical modelling which uses combat theory as an
       external control on combat modelling;

Soft Analysis Inside Studies (Saaty Method)

(5)   identify and eventually concentrate on the principal level
       within the hierarchy of operations;
(6)   treat data parametrically, using plausible ranges rather
       than precise values;
(7)   examine extreme cases as well as particular cases of
       specific interest;
(8)   reduce the number of choices and attributes to the
       absolute minimum;
(9)   concentrate on the pairwise comparisons of choices
       throughout;

Combat Theory Inside Studies (Lanchester Theory)

(10)   use combat theory to give the trends and sensitivities of
       the big picture;
(11)   look for existing theoretical results which have a direct
       link with the required analytical judgement;
(12)   use combat theory to plan the combat modelling so as to
       get the most needed information for the least effort;
(13)   concentrate on pairwise comparisons, extreme cases and
       first-order effects, in line with suggestions (9), (7) and
       (2) respectively;
(14)   sustain the algebraic manipulation of symbols;
(15)   concentrate at first on the deterministic combat power
       function and the equation of state;
(16)   concentrate on strategies and tactics which lead to the
       separation of the combat powers of the two sides within
       the equation of state;
(17)   introduce termination conditions and then the combat power
       advantage and the probability of victory;
(18)   concentrate on the condition of parity between combatants,
       in the forms of deterministic parity and stochastic power
       parity and probability parity;

Table 4 (continued)

(19) make extensive use of <u>non-dimensional parameters</u>;
(20) introduce non-dimensional forms of the threat, or attack,
      and the deterrent, or defence, and call them the Attack
      parameter and the Defence parameter;
(21) call the ratio Defence parameter/Attack parameter the
      Lanchester number;
(22) find the value of the Lanchester number under the
      condition of deterministic parity, and divide the
      Lanchester number by its parity value;
(23) call the resultant non-dimensional parameter the Fiske
      number, and regard it as the pivotal parameter of the
      problem from the planning point of view;

Combat Modelling Inside Studies (Simulation Modelling)

(24) use combat modelling for numerical experiments to
      illustrate detailed features of the problem already
      treated approximately by combat theory;
(25) use a well-documented and transparent simulation model;
(26) prefer a simulation model in which theory gives a measure
      of internal control;
(27) concentrate on expected values and variances at most;
(28) use a variance reduction technique to calculate the combat
      invariant from which the stochastic equation of state may
      be determined;
(29) concentrate on the <u>parity condition</u> in its three forms;
(30) analyse the parity condition by means of the Fiske number;
(31) plot the probability of victory against the Fiske number.

Combat Theory Outside Studies

(32) develop combat dynamics as if it were a physical science
      such as aeronautical fluid dynamics;
(33) take generalised Lanchester theory as the formal basis of
      combat theory and correlate with the Dupuy model of land
      combat;
(34) consolidate results from numerical experiments done by
      combat modelling, for instance by system dynamics
      techniques;
(35) combine theory and the results of modelling into a
      permanent body of quantitative information in the form
      of a manual for use in Studies.

6.2 Combat Theory (Lanchester Theory)

6.2.1 Practical Roles

The applications of combat theory in Studies may be divided into two classes. First there are those in which combat theory contributes building blocks to the combat modelling, thus providing a kind of internal control as part of the process of validation. This class contains the main application of combat theory at the present time; fòrmulae from Lanchester theory are particularly prominent as the basis of certain types of simulation modelling. The second class includes those uses in which combat theory exerts a form of external control over combat modelling. I call these the practical roles of combat theory.

They may themselves be divided into three types: (i) those which exert control within the model world but are not, as it were, hidden inside the modelling; (ii) those which in some way effect a transition between the model world and the real world by shedding light on what it might be feasible and reasonable to attempt through combat modelling, bearing in mind the use to which the results are to be put; (iii) those which apply directly to the real world in that the theoretical results could be used as they stand to help in forming the balanced analytical judgement. Examples from the work of myself and colleagues at DOAE are given in the References (see Table 5); Example 6 in particular applies many of the theoretical ideas put forward in this paper.

From the point of view of the progress of a Study, the three types of practical roles of combat theory are best considered in the reverse order. The first thing is to unearth from existing theory, or derive from basic principles, results which provide an immediate link with the burgeoning analytical judgement. The next requirement is to use the theory to plan the combat modelling so as to get the most information for the least effort. Then the theory should be put in the precise form which is most suitable for the presentation of the modelling results.

The extent to which existing combat theory will contribute to the analysis in a Study will depend closely on the Study question. At best, it will give a clear indication of the overall numerical sensitivities of the problem and show where detailed examination by simulation modelling is required. Even if theory does not lead to fresh insights, it might well provide a compact reminder of old lessons that might otherwise be forgotten. But, at worst, there may be no suitable theory at all. In this case it will be up to the Study team to derive the

Table 5   References to Personal Work on the Practical Roles of
Combat Theory.

Practical Roles		Examples			
Type	Description	No.	Topic	Reference	Comment
(i)	Exert Control Within Model World	1.	Air Defence of Ships	[61], Example I (with J.A. Hall)	Co-ordinates Theory with Simulation Modelling
		2.	Mixed-Fire Land Combat	[60]	Compares Different Types of Combat
(ii)	Provide Transition from Model to Real World	3.	Direct-Fire Land Combat	[61], Example II (with T.G Weale)	Compares Different Analytical Treatments
		4.	Air Support of Land Battle	[3] (with R.G. Body)	Shows Relative Influence of Air and Land Operations
(iii)	Apply Directly to Real World	5.	Suppression of Air Defences	[59] (with D. George)	Demonstrates Robust Tactical Option
		6.	Air Defence of Primary Targets	[60] [61], Example III (with B.M.E. Rodgers)	Displays Cost-Effective Planning Option

required theory for themselves, if they have the necessary skill
and enough time at their disposal. Their aim should be to
produce the traditional deterministic back-of-the-envelope
model. The rest of this sub-section describes some measures
designed to help a Study team to make appropriate extracts
from existing theory or derive their own.

6.2.2 Suggested Measures

If, as I suggest, Studies require only gross quantitative
outputs, then the theory that provides part of the analytical
experience need only be a first-order theory. There may be an
exception here to the general rule (Bondi [5]) of the
inadequacy of simple mathematics in real life. If military
judgement is the arbiter, and the pursuit of all-embracing
verisimilitude does not work, perhaps selective simplicity
(applied to the real problem through the judgemental analysis)
might pay off. Consider therefore some of the things a Study
team might look for when trying to be selective and to simplify
at the same time.

By combining the features of the Saaty method with those of
Lanchester theory described earlier, and bearing in mind the
requirements of simulation modelling, I am led to suggest that
it might be worthwhile for a Study team to:

(1) concentrate on pairwise comparisons of the minimum number
     of choices;
(2) examine the extreme cases which bound the Study question;
(3) consider only first-order effects;
(4) sustain the algebraic manipulation of symbols;
(5) concentrate at first on combat effectiveness as expressed
     through the deterministic combat power function and the
     equation of state;
(6) concentrate on strategies and tactics which lead to
     equations of state in which the combat powers of the two
     sides are separable;
(7) introduce termination conditions, and the combat power
     advantage, and in stochastic treatments consider the
     probability of victory as a second expression of combat
     effectiveness;
(8) concentrate on the parity condition in the forms of
     deterministic parity, power parity and probability parity;
(9) make extensive use of non-dimensional parameters.

These various measures, which are included in Table 4, are
self-consistent and should be applied together for maximum
effect. Non-dimensional parameters are of particular interest
and are highlighted in Table 3.

6.2.3 Non-dimensional Parameters

Non-dimensional parameters are perhaps of greater use than
any other single feature of combat theory. In the previous
section I have described certain types of non-dimensional
parameters and ventured to give them popular names. However,
it matters less what they are called than that the parameters
and their names should be widely understood and used. To
repeat briefly, the parameters are of three types. The first
type consists of the Attack (or Threat) parameter and the
Defence (or Deterrent) parameter. The second type is
exemplified by the ratio Defence parameter/Attack parameter,
which I suggest might be called the Lanchester number. The
third type is the ratio of the Lanchester number and its
parity value, which I suggest might be called the Fiske
number. The Fiske number is an important parameter for the
purposes of planning.

If, in the absence of an existing theory, a Study team has
not got the time and skill to produce a back-of-the-envelope
model, at least they should deduce the dominant non-dimensional
parameters of the problem. The first step should be to
determine the two most significant non-dimensional input
parameters, namely the Attack and Defence parameters. The
Lanchester number would then follow, but without a theoretical
solution to the problem it would not be possible to determine
a parity condition and so define the Fiske number. The next
step could be to define two, at most, non-dimensional output
parameters; one of them might be the probability of victory,
the other should be one that could be derived from a
deterministic analysis in principle. Even without a connecting
theory the non-dimensional input and output parameters could
be used to plan the combat modelling and to analyse its results,
either independently or empirically.

6.3 Combat Modelling (Simulation Modelling)

6.3.1 Controlled Modelling

The measures outlined above could help to impart a definite
structure to the analysis attempted in a Study. They should be
applied in a way which enables the great strengths of simulation
modelling to be utilised to the full. Simulation modelling can
usually be relied upon to give an answer to complex problems
well beyond the reach of symbolic analysis. It can cope with
problems which do not readily exhibit any discernible pattern.
And it can give an appearance of realism to the modelling
process. The members of a Study team can derive much comfort
from the possession of detailed simulation models. But

numerical experiments are their forte, and to provide reliable
numerical experiments simulation models require two forms of
control, one internal, the other external.

Internal control concerns the validity of the results
generated by the modelling. Validity is best demonstrated by
models which are well-documented and clear and comprehensible,
even though complex. Simulation models based on the processes
of stochastic Lanchester theory, such as those developed in the
UK by Haysman and his colleagues [39,68,74] , have a high
potential for internal control. So has the model recently
described by Sassenfeld [78] , which is almost as much a part
of combat theory as of combat modelling. The models SLEW (of
Haysman and colleagues [74]) and ELAN (of Sassenfeld [78]) are
ideally suited for external control by combat theory.

There are many ways in which symbols and simulations could
be compared or combined. (Cockram [15] and Grainger [32]
describe two early successful examples from work done at DOAE.)
What I am suggesting is, in the first place, simply a loose
association in which theoretical results form a backcloth
against which the results of modelling are viewed. The two
sets of results would be independent, except for the influence
of the theory on, initially, the planning of the modelling
runs and, subsequently, the presentation of the results. This
sort of method, applied to theory and physical experiments,
has been standard throughout the history of aeronautical fluid
dynamics (for instance, see Goldstein [31] and Howarth [43]).
It is used today, with the same basic symbolism, with respect
to numerical experiments (see Green [33] and in particular
Lock and Firmin [57]). For the Lanchester theory of combat
dynamics to be used in defence OA in a manner similar to the
use of the theory of fluid dynamics in aeronautics would be
especially appropriate since Lanchester himself was also a
renowned fluid dynamicist and aeronautical engineer (see von
Karman [49]).

If theory, whether already existing or specially derived,
could supply the input and output non-dimensional parameters
and provide functional relationships between them, it would
become the driving factor in the analysis. The combined
method might then be called empirically enhanced symbolic
analysis (or EESA for short?). It could be a good way of
exerting external control on simulation modelling through the
parity condition. This feature, which is emphasised by
Sassenfeld [78] , is given prominence in Table 3 and discussed
further below.

6.3.2 Parity Condition

The parity condition concerns expected values. Expected values and variances are widely considered to be the least that should be obtained from a stochastic simulation model. But, in my experience, they are the very most that can be digested by a Study team in a time-limited Study. Usually, a Study team is thankful just to be able to understand the expected values. There is much scope for variance reduction techniques which enable expected values to be derived more economically than by the averaging of hundreds of replications of each case. To minimise the number of runs needed to analyse power parity in a Study, it would be extremely useful to have a martingale technique of variance reduction which led directly to the combat invariant from which the stochastic equation of state and the combat power advantage could be determined. Probability parity could be examined at the same time by inspection of the outputs for the probability of victory.

A way of presenting the results of simulation modelling, derived from my own experience, is illustrated diagrammatically in Figure 4. The defender's probability of victory is plotted against the (defender's, by definition) Fiske number for a fixed value of a secondary (input) non-dimensional parameter. The Fiske number is deduced from simplified deterministic theory and used to select cases for which its value is near unity. Stochastic power parity occurs well within the range of cases drawn in Figure 4. Probability parity is also present. (Deterministic parity, power parity and probability parity are usually much closer than shown; often they are indistinguishable.) The gradient of the probability curve at the parity point varies with changes in the secondary non-dimensional parameter. It could be made, together with the parity point, the basis of an approximation that would be sufficient for practical analytical purposes. Such a form of analysis, which is listed in Table 4, might have a unifying effect if adopted in Studies. But the greatest unifying and instructive influence would be an accepted theory of combat dynamics built up outside Studies.

PROBABILITY OF VICTORY

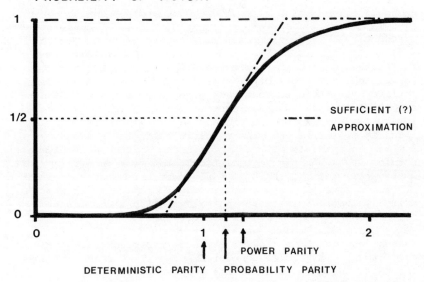

Fig. 4 Presentation of Results of Combat Modelling

7. COMBAT THEORY OUTSIDE STUDIES

It has been stressed throughout this paper that the quantitative analysis to be done in a Study would be easier and more convincing if it were based on and allied to a coherent and comprehensive theory of combat. The MOD appears not to have recognised the need for such a theory in the past. The reason may lie in the resources required to do a complete job, to supply the necessary analytical effort and the outstanding knowledge and skill. It is surprising, however, in view of the close links at the highest levels between the MOD and the Institute of Mathematics and its Applications, in particular through the Defence Scientific Advisory Council (DSAC), that something substantial has not been done.*

The Military Conflict Institute in the USA (described by Marshall [65]), an independent association of 'soldiers and scholars, engineers and operations analysts, historians and scientists', has set itself the daunting task of producing a book on 'A Theory of Combat' by the middle of 1987. The aim is to produce a first-cut statement, in words, of a theory of combat for use in developing effective models and plans, in training, and in decision-making. The theory will be designed

---

* Note from the IMA - The links are not official. Several
  Fellows of the Institute hold responsible positions in the
  Advisory Council.

to be of direct value to a wide spectrum of potential users.
It will be derived in part from the combat model constructed
by Dupuy [23] , and will have many expert contributors.

The theory of combat being developed by the Military Conflict
Institute derives principally from the direction of history and
philosophy. The word theory is given a more general meaning
than the specialized usage adopted in this paper. In the
particular context of the present paper, the theory is
intended to fulfil the primary practical role of a theory of
combat, that is, to be capable of being applied directly in
forming the quantitative analytical judgement in a Study. As
a practical manual, the book could turn the hypothetical extreme
of Figure 3d into a real possibility. Certainly, if the
Military Conflict Institute succeeds in its endeavour all
defence operational analysts will be in its debt.

From the point of view of the Study procedure proposed in
this paper, with its content of empirically enhanced symbolic
analysis, there might be a case for a supplementary book.
(Another book? - as if producing even one book on the theory
of combat were not difficult enough!) This second book would
concentrate on the two other practical roles of combat theory,
those which exert control over combat modelling outside the
model but within the model world, and those which provide a
transition between the model world and the real world by
displaying the numerical sensitivities of a Study question. It
would aim to elucidate what might be called the problem of known
sensitivity, as a corollary of the problem of uncertainty.
While the many-faceted problem of uncertainty is so huge that
it is unlikely ever to be resolved in its entirety, the problem
of known sensitivity could be treated to a useful extent. It
is known, for instance, that factors of two can mean the
difference between victory and defeat, and factors of two are
everywhere in combat dynamics. A suitable book could address
the problems of sensitivity, scale and criteria with some hope
of showing in advance what might happen in defined circumstances
of outstanding interest.

Such a book would require the development of combat dynamics
as if it were a physical science (see Table 4); the book itself
might resemble the celebrated volumes on aeronautical fluid
dynamics edited by Goldstein [31] nearly half a century ago.
Lanchester theory could provide the formal basis, starting
from the treatise by Taylor [86] , and be correlated with the
land combat model of Dupuy [23] . It could be augmented by
sanitised results of numerical experiments, consolidated (by
techniques like that of Farrell [28] or of system dynamics)
and made scenario-independent. The material in the book should
be chosen so as to be of maximum use in Studies. Compiling such

a manual would be a most arduous undertaking. Over a thousand
man years of effort, appropriate to the establishment of a
science of combat dynamics, have been lost since the Second
World War in the UK alone. Perhaps only the inspired use of
the capabilities of modern computers to perform symbolic
analysis could help to make up for the lost opportunity.

8. CONCLUDING SUMMARY

In this paper I have tried to give a fair and constructive
critique of the mathematical modelling of combat in defence
OA in the UK. At the same time, I have followed the particular
theme of Lanchester theory in the practice of defence Studies.
Although some of the criticisms I have recorded are harsh, I
am entirely sympathetic to defence operational analysis as a
subject. The years I spent at DOAE were among the most
stimulating and enjoyable of my career, and my wish in writing
this paper is to assist the cause of the quantitative analysis
of defence operations. The remarks which follow, when taken
with the Glossary and Tables 1-4, give a largely self-contained
summary of the main points of each section of the paper.

Defence OA may be surveyed briefly as follows. (i) It is
a vast, complex, rather strange and very difficult subject.
(ii) It covers all the activities of the Armed Services, with
the emphasis on the dynamics of combat. (iii) By definition,
it is an aid to decision-making based on quantitative analysis.
(iv) It requires the forming of analytical judgement from
quantitative analytical experience. (v) Forming analytical
judgement involves soft analysis, of which the Saaty method is
an example. (vi) Quantitative analytical experience of
combat dynamics is gained by hard analysis, principally through
the mathematical modelling of combat. (vii) The mathematical
modelling of combat consists of symbolic combat theory and
numerical combat modelling, that is, the doing of numerical
experiments. (viii) The main form of combat theory is
Lanchester theory and the main form of combat modelling is
simulation modelling. (ix) Defence OA is dominated by time-
limited Studies on specific questions, sponsored by the Armed
Services to assist the making of planning decisions. (x) The
practice of defence Studies is dominated in turn by combat
simulation modelling.

Combat simulation modelling in Studies encounters formidable
problems. It suffers from: (a) the complexity of simulation
models; (b) the lack of good data; (c) the difficulties
associated with uncertainty, scale and the absence of agreed
criteria of assessment; (d) the impracticality of doing more
than a few numerical experiments in one Study. These problems
are compounded by the critical omissions which are a consequence

of the way defence OA is conducted. They are: (i) soft
analysis such as the Saaty method is only rarely used in
Studies; (ii) the results from numerical experiments by
combat simulation modelling are not usually collected together
for subsequent use; (iii) the available combat theory is
scarcely ever used directly in Studies, the chief role of
Lanchester theory being as the basis of certain types of
simulation modelling; (iv) little effort is devoted to any
analysis of combat dynamics outside Studies, with the result
that the theoretical material accumulated in the past has not
been developed into an accepted general theory of combat.
Defence OA is therefore made to seem even more difficult than
it is, and, without a firm quantitative foundation, analytical
judgements are liable to be formed from ad hoc analytical
experience gained under the urgent pressures of expediency.

I think that an attempt to repair the above omissions
would have a beneficial effect on the problems of combat
simulation modelling. There seems to be a case for a
different approach to the practice of defence Studies. This
paper suggests an approach which differs from the present
approach in three respects by proposing: (i) an attitude of
mind which accepts that analytical judgement must ultimately
be subordinate to military judgement and which seeks only
gross quantitative statements of analytical experience; (ii)
a mixture of analytical methods which includes more soft
analysis, more combat theory and less combat modelling; (iii)
a procedure which separates the soft cover of analytical
judgement from a hard core of quantitative analytical experience,
and combines combat theory and combat modelling into an
empirical form of the mathematical modelling of combat.

These general suggestions should benefit the overall quality
of the analysis done in Studies. However, it is also
necessary to try to produce a high standard of work in each
analytical component. As an introduction to the description of
some more-detailed measures with this object in view, I outline
some features of Lanchester theory which are basic to the
detailed development of the general approach. They include:
the pattern of solution for the course of a battle, involving
the initial conditions, the equation of state and the
termination conditions; two measures of the combat effectiveness
of a force (the combat power advantage and the probability of
victory); the parity condition, which determines when two
forces are evenly matched; and non-dimensional parameters which
help to simplify analysis and increase analytical understanding.

The ideas in the general approach are then made specific by
considering the Saaty method, Lanchester theory and simulation

modelling as special cases of soft analysis, combat theory and combat modelling respectively. The main detailed suggestions are listed in Table 4. The primary suggestions, underlined in Table 4, are for concentration on the pairwise comparisons of choices, extensive use of non-dimensional parameters, and concentration on the parity condition.

While Table 4 contains quite a long list of suggestions, most of them are really very elementary. Their importance lies in their having been largely neglected in the past in the UK. But the greatest need is for a coherent and comprehensive theory of combat dynamics, eventually combined with the results of numerical experiments by simulation modelling, to be built up outside Studies in such a way that it becomes an easily accessible basis for forming analytical judgements. So long as there is this void, and only the Military Conflict Institute in the USA seems to have the courage to try to fill it, analytical judgements in defence OA will continue to be subject to suspicion.

ACKNOWLEDGEMENT

This paper is based on the author's official experience and is published with the permission of the Director, Defence Operational Analysis Establishment. Copyright ⊙ Controller HMSO, London, 1986.

REFERENCES

Note. An asterisk indicates a reference included for its interest and general relevance but not mentioned specifically in the text.

1. Andrews, D.R. and Laing, G.J., (1986) 'Some Problems of Modelling Battle', this volume.

2. Blumenthal, D.K., (1984) 'Bridging the Gap between Wargames and Closed Simulations', Lawrence Livermore National Laboratory Preprint UCRL-91298, International Symposium on Advances in Combat Modelling, Shrivenham.

3. Body, R.G. and Lord, W.T., (1975) 'An Elementary Examination of the Influence of an Air Campaign on a Land Battle', DOAE Working Paper No. (Air) 120 (Unpublished MOD report).

4. Bonder, S., (1970) 'Systems Analysis: a Purely Intellectual Activity', Keynote address to 25th Military OR Symposium, New London; also (1971) Military Review Vol. 51, No. 2, pp. 14-23.

5. Bondi, Sir Hermann, (1985) 'The Applicability of Mathematics',
    IMA Bulletin, Vol. 21, Nos. 3/4, pp. 46-47.

6. Bowen, K.C., (1967) 'A Lanchester-Type Stochastic Model
    Related to the Operational Evaluation of ASW Weapon
    Systems', DOAE Accession No. 34383, NATO Conference on
    Recent Developments in Lanchester Theory, Munich.

7*. Bowen, K.C., (1972) 'The Scope for Mathematics in
    Operational Research', IMA Bulletin, Vol. 8, No. 3, pp. 78-
    84.

8. Bowen, K.C., (1973) 'Mathematical Battles', IMA Bulletin,
    Vol. 9, No. 10, pp. 310-315.

9. Brackney, H., (1959) 'The Dynamics of Military Combat',
    Opns. Res. Vol. 7, No. 1, pp. 30-44.

10. Callahan, L.G. Jr. and Gagliano, R.A., (1984) 'Weapon
    Systems and Combat Models: Their Evolution and Functional
    Regularities', Georgia Institute of Technology, International
    Symposium on Advances in Combat Modelling, Shrivenham.

11.* Checkland, P.B., (1983) 'OR and the Systems Movement:
    Mappings and Conflicts', J. Opl. Res. Soc. Vol. 34, No. 8,
    pp. 661-675.

12. Chief Scientific Adviser MOD (1978), 'OA in Defence',
    MOD Seminar with Universities and Industry, Sunningdale,
    particularly papers 'Problem of Uncertainty', by G.H.B.
    Jordan, 'Problem of Scale' by B.A.P. James and 'Problem of
    Criteria'by P.M. Sutcliffe; Unpublished MOD papers.

13.* von Clausewitz, C., (1832), 'On War', edited and with an
    introduction by A. Rapoport (1968), Pelican Books,
    Harmondsworth, Middlesex.

14. Cockle, A.A.V., (1985) 'Qualitative Assessment BAMUSE',
    British Aerosapce, Second International Symposium on
    Military Operational Research, Shrivenham.

15. Cockram, R., (1972) 'The Use of Lanchester's Equations
    to Predict the Degree of Disruption of Ground Attack
    Missions by Fighter Patrols', DOAE Working Paper No.
    (Air) 47 (Unpublished MOD report).

16. Comptroller General of the United States Report to
    Congress (1980), 'Models, Data and War: a Critique of the
    Foundation for Defense Analyses', US General Accounting
    Office PAD-80-21.

17. Coyle, R.G., (1981) 'A Model of the Dynamics of the Third
     World War - An Exercise in Technology Transfer', *J. Opl
     Res. Soc. Vol. 32,* No. 9, pp. 755-765.

18. Daly, F., (1982) 'Lanchester Theory of Conflict,
     References', UK Lanchester Study Group, c/o Department of
     Statistics and OR, Faculty of Applied Science, Lanchester
     Polytechnic, Coventry.

19. Daly, F., (1985) 'The UK Lanchester Study Group', in
     Special Issue on Management Science in Defence edited by
     Bowen, K.C., OMEGA, Vol. 13, No. 2, pp. 131-133.

20.* Daniel, D.W., (1985) 'The Politics, Philosophy and
     Practice of OR - A Personal View of the Issues', in Special
     Issue on Management Science in Defence edited by Bowen, K.C.,
     OMEGA, Vol. 13, No. 2, pp. 89-94.

21.* Daniel, D.W. and Dare, D.P., (1983) 'Cost and Utility:
     Where does the Balance Lie for Governments', *J. Opl Res.
     Soc. Vol. 34,* No. 3, pp. 193-200.

22. Director DOAE (1979) 'Procedure for Conducting Studies
     at DOAE' Unpublished MOD Memorandum.

23. Dupuy, T.N., (1979) 'Numbers, Predictions and War',
     MacDonald and Jane's, London.

24. Edwards, J., (1984) 'The Application of Lanchester's
     Equations to Naval EW Assessment', MEL, International
     Symposium on Advances in Combat Modelling, Shrivenham.

25.* Eilon, S., (1982) Address following Presentation of Silver
     Medal of Operational Research Society, *J. Opl Res. Soc.
     Vol. 33,* No. 12, pp. 1091-1098.

26. Engel, J.H., (1963), 'Comments on a Paper by H.K. Weiss',
     Opns. Res. Vol. 11, No. 1, pp. 147-150.

27. Everett, A.F., Harmsworth, B., Miller, R.J.R. and
     Hanbury, R., (1985) 'An Analysis of Tank Attributes', MOD
     and EASAMS, Second International Symposium on Military
     Operational Research, Shrivenham.

28. Farrell, R.L., (1984) 'MACRO: a New Development in Large-
     Scale Combat Modelling', Vector Research Incorporated,
     International Symposium on Advances in Combat Modelling,
     Shrivenham.

29. Forrester, J.W., (1971) 'World Dynamics', Wright-Allen
     Press Inc., Cambridge, Massachusetts.

30. Godfrey, L.D., and Anderson, M.R., (1985) 'The Use of the
     Analytic Hierarchy Process for Prioritizing Materiel
     Developments and Procurements', U.S. Army Combined Arms
     Center and U.S. Army Combined Arms Operations Research
     Activity, Second International Symposium on Military
     Operational Research, Shrivenham.

31. Goldstein, S., (Editor) (1938) 'Modern Developments in
     Fluid Dynamics', Volumes I and II, Oxford University Press,
     Oxford.

32. Grainger, P.L., (1976) 'The Use of a Stochastic Tank
     Engagement Model to Examine the Effects of Moving Target
     Correlation and to Derive Equivalent Lanchester Equations',
     MSc (OR) Thesis, Brunel University.

33. Green, J.E., (1982) 'Numerical Methods in Aeronautical
     Fluid Dynamics - An Introduction', in 'Numerical Methods in
     Aeronautical Fluid Dynamics', Editor Roe, P.L., pp. 1-32,
     Academic Press, London.

34. Gye, R. and Lewis, T., (1976) 'Lanchester's Equations:
     Mathematics and the Art of War - A Historical Survey and
     Some New Results', *Math. Scientist, Vol. 1,* pp. 107-119.

35. Hartley, D.A., Hagues, J.N. and Kettle-White, W., (1981)
     'Deterministic and Stochastic Lanchester's Equations: A
     Comparison using Simulation Techniques', RMCS Department of
     Management Sciences Report OR/WP/12.

36. Hartley, D.A. and Haysman, P.J., (1985) 'The Use of
     Judgemental Methods in Defence Operational Research', Royal
     Ordnance, Second International Symposium on Military
     Operational Research, Shrivenham.

37. Haysman, P.J., and Mortagy, B.E., (1980) 'References on
     the Lanchester Theory of Combat to 1980', RMCS Department
     of Management Sciences Working Paper OR/WP/6.

38. Haysman, P.J., (1980) 'Formalising Subjective Judgements',
     RMCS Department of Management Sciences, Internal Notes on
     Operational Analysis, OA 285.

39. Haysman, P.J. and Wand, K., (1986) 'STOIC: A Method for
     Obtaining Approximate Solutions to Heterogeneous Lanchester
     Models', this volume.

40. Helmbold, R.L., (1985) 'Combat History Analysis Study Effort (CHASE): A Progress Report', U.S. Army Concepts Analysis Agency, Second International Symposium on Military Operational Research, Shrivenham.

41.* Hollingdale, S.H., (1978) 'Methods of Operational Analysis', in 'Newer Uses of Mathematics', Editor Sir James Lighthill, pp. 176-280, Penguin Books Ltd., Harmondsworth, Middlesex.

42. Hough, B.F., and Isbrandt, S., (1985) 'Methods for Evaluating Contractor Proposals', ORAE Canada, Second International Symposium on Military Operational Research, Shrivenham.

43. Howarth, L., (Editor) (1953) 'Modern Developments in Fluid Dynamics - High Speed Flow', Volumes I and II, Oxford University Press, Oxford.

44.* Huber, R.K., (1985) 'On Current Issues in Defence Systems Analysis and Combat Modelling', in Special Issue on Management Science in Defence edited by Bowen, K.C., OMEGA, Vol. 13, No. 2, pp. 95-106.

45. Hughes, W.P., Jr (Editor)(1984) 'Military Modeling', Military Operations Research Society, Alexandria, Virginia.

46. Hynd, W.R.B., (1982) 'Studies in Lanchester Theory by the Cunningham School, 1937 Onwards', Rex, Thompson & Partners Tech. Report No. RTP/2590/01.

47. James, B.A.P., (1981) 'A Random Walk through Lanchester Square', Unpublished MOD report (DOAE Working Paper 37/7 (2/81)).

48. Jones, R.V. (1982) 'A Concurrence in Learning and Arms', J. Opl Res. Soc. Vol. 33, No. 9, pp. 779-791.

49. von Karman, T., (1958) 'Lanchester's Contributions to the Theory of Flight and Operational Research', JR Aero Soc., Vol. 62, No. 2, pp. 80-93.

50. Karr, A.F., (1981) 'Lanchester Attrition Processes and Theater Level Combat Models', Institute for Defense Analyses Paper P-1528; also (1983) in 'Mathematics of Conflict', Editor Shubik, M., pp. 89-126, North Holland, Elsevier Science Publishers B.V., Amsterdam.

51.* Kerr, T.H., (1969) 'The Role of Operational Analysis in Defence Studies', RUSI Journal, Vol. 114, No. 1, pp. 41-52.

52. Koehler, H., (1984)   'Combat Representation in the
     Computer-Based War Game', AStudUbBw West Germany,
     International Symposium on Advances in Combat Modelling,
     Shrivenham.

53. Koopman, B.O., (1964) 'Analytical Treatment of a War
     Game', in 'Proceedings of the Third International
     Conference on Operational Research (Oslo 1963)', Editors
     Kreweras G., and Morlat, G., pp. 727-735, English
     Universities Press, London.

54. Laing, G.J. (1986)   'The Mathematical Equivalence of
     Conceptual Diagrams and the Use of Electronic Worksheets
     for Dynamic Models', this volume.

55. Lanchester F.W., (1916) 'Aircraft in Warfare: The Dawn
     of the Fourth Arm', Constable and Co. Ltd., London.

56.* Lawrence, E.G., (1984)   'Some Experiences in the Teaching
     of Mathematical Modelling', IMA Bulletin Vol. 20, Nos. 7/8,
     pp. 121-123.

57. Lock, R.C. and Firmin, M.C.P., (1982)   'Survey of
     Techniques for Estimating Viscous Effects in External
     Aerodynamics' in 'Numerical Methods in Aeronautical
     Fluid Dynamics', Editor Roe, P.L., pp. 337-430.

58. Lockett, A.G., Muhlemann, A.P. and Gear, A.E., (1981)
     'Group Decision Making and Multiple Criteria - A Documented
     Application', in 'Proceedings of the Fourth International
     Conference on Multiple Criteria Decision Making (Newark,
     Del., U.S.A.)', Edited by Morse, J.N., pp. 205-221.
     Springer-Verlag, New York.

59. Lord, W.T. and George, D., (1977)   'An Analytical Model
     of Defence Suppression Operations', DOAE Memorandum 76103.

60. Lord, W.T., (1977)   'Simple Mathematical Analysis as a
     Preliminary to Modelling', Unpublished MOD report (DOAE
     Working Paper 615/1).

61. Lord, W.T., (1984a)   'Combat Modelling in Defence Studies',
     Rex, Thompson & Partners, International Symposium on
     Advances in Combat Modelling, Shrivenham; also RTP Document
     RTP/030/03/01.

62. Lord, W.T., (1984b), 'Lanchester Theory in Practice',
     Rex, Thompson & Partners, IMA Conference on the Mathematical
     Modelling of Conflict and its Resolution, Cambridge; also
     RTP Document RTP/030/03/02.

63. Low, L.J., (1981) 'Theater-Level Gaming and Analysis
     Workshop for Force Planning, Volume II - Summary, Discussion
     of Issues and Requirements for Research', SRI International.

64.* Majone, G. and Quade, E.S., (Editors) (1980), 'Pitfalls
     of Analysis', Volume 8 of International Series on Applied
     Systems Analysis, Wiley, Chichester.

65. Marshall, D.S., (1984) 'The Military Conflict Institute -
     Purpose, Methodology and Results to Date', Military Conflict
     Institute, International Symposium on Advances in Combat
     Modelling, Shrivenham.

66. Mertes, E., (1984) 'Tactics and Terrain - Interdependency
     and Transferability of Combat Results', Industrieanlagen -
     Betriebsgesellschaft mbH, International Symposium on
     Advances in Combat Modelling, Shrivenham.

67. Morse, P.M. and Kimball, G.E., (1951) 'Methods of
     Operations Research', The MIT Press, Cambridge,
     Massachusetts and Wiley, New York.

68. Mortagy, B.E., (1981) "Extensions to the Lanchester
     Model of Combat for the Analysis of Mixed Force Battles',
     PhD Thesis, Operational Research Branch, RMCS.

69.* Müller-Merbach, H., (1984) 'Interdisciplinarity in
     Operational Research - in the Past and in the Future',
     J. Opl Res. Soc., Vol. 35, No. 2, pp. 83-89.

70. Nield, R.R., (1986) 'Accuracy and Lanchester's Law: A
     Case for Dispersed Defence?', this volume.

71.* Pitt, Sir Harry,(1985) 'The Place of Mathematics', IMA
     Bulletin, Vol. 21, Nos. 3/4., pp. 61-65.

72.* Polya, G., (1957) 'How to Solve It', Second Edition
     Princeton University Press, Princeton; also (1971) Princeton
     Paperback, First Printing.

73. Price, T., (1986) 'The Role of Operational Analysis',
     this volume.

74. Read, M.J., Wand, K. and Haysman, P.J., (1984) 'SLEW MK1:
     A Direct-Fire Model at Battle Group Level', RMCS,
     International Symposium on Advances in Combat Modelling,
     Shrivenham.

75.* Rivett, B.H.P., (1978) 'Planning' in 'Newer Uses of
      Mathematics', Editor Sir James Lighthill, pp. 383-428,
      Penguin Books Ltd., Harmondsworth, Middlesex.

76.   Rowland, D., (1984) 'Field Trials and Modelling', DOAE,
      International Symposium on Advances in Combat Modelling,
      Shrivenham.

77.   Saaty, T.L., (1980) 'The Analytic Hierarchy Process',
      McGraw-Hill, New York.

78.   Sassenfeld, H.M., (1985) 'Battalion Level Combat
      Analysis with the Model ELAN', U.S. Army TRADOC Systems
      Analysis Activity, Second International Symposium on
      Military Operational Research, Shrivenham.

79.* Shephard, R.W., (1983) 'Resource Allocation in Defence',
      RMCS Department of Management Sciences, Internal Notes on
      Operational Analysis, OA 861 (Issue 2), presented at ORS
      25th Annual Conference, Warwick.

80.   Shubik, M. and Brewer, G.D., (1972) 'Models, Simulations
      and Games - A Survey', RAND R-1060-ARPA/RC.

81.   Shubik, M., (Editor) (1983) 'Mathematics of Conflict',
      North Holland, Elsevier Science Publishers B.V., Amsterdam.

82.   Stockfisch, J.A., (1975) 'Models, Data and War: A
      Critique of the Study of Conventional Forces', RAND R-1526-
      PR.

83.* Stratton, A., (1968) 'The Principles and Objectives of
      Cost Effectiveness Analyses', *Aeronautical Journal,* Vol. 72,
      No. 685, pp . 43-53.

84.   Strauch, R.E., (1974) 'A Critical Assessment of
      Quantitative Methodology as a Policy Analysis Tool', RAND
      P-5282; also (1983) in 'Mathematics of Conflict', Editor
      Shubik, M., pp. 29-54, North Holland, Elsevier Science
      Publishers B.V., Amsterdam.

85.* Sun Tzu (400BC?), 'The Art of War'; (1963), translated
      and with an introduction by S.B. Griffiths, Oxford
      University Press, London; (1971), OUP Paperback.

86.   Taylor, J.G., (1983) 'Lanchester Models of Warfare',
      Volumes I and II, Military Applications Section, Operations
      Research Society of America, c/o Ketron Inc., Arlington,
      Virginia.

87.* Tomlinson, R. and Kiss, I., (Editors) (1983) 'Rethinking the Process of Operational Research and Systems Analysis', Volume 2 of Frontiers of Operational Research and Applied Systems Analysis, Pergamon Press Ltd., Oxford.

88. Watson, R.K., (1976) 'An Application of Martingale Methods to Conflict Models, Opns.Res. Vol. 24, No. 2, pp. 380-382.

89. Weale, T.G., (1971) 'The Mathematics of Battle I: A Bivariate Probability Distribution'; DOAE Memorandum 7129.

90. Weale, T.G., (1972) 'The Mathematics of Battle II: The Moments of the Distributions of Battle States', DOAE Memorandum 7130.

91. Weale, T.G., (1975) 'The Mathematics of Battle V: Homogeneous Battles with General Attrition Function', DOAE Memorandum 7511.

92. Weale, T.G., et al. (1973-1978) Further analyses of the Mathematics of Battle, III, IV, VI, VII, VIII, DOAE Memoranda 7315, 7316, 76126, 77105, 78106; also (1979) 'Stochastic Lanchester Theory using Numerical Methods', ORS Annual Conference, Stirling.

93. Weiss, H.K., (1953) 'Requirements for a Theory of Combat', Ballistic Research Labs., Aberdeen Proving Ground, Maryland, BRL Report No. 667; also (1983) in 'Mathematics of Conflict', Editor Shubik, M., pp. 73-88, North Holland, Elsevier Science Publishers B.V., Amsterdam.

94. Weiss, H.K., (1962) 'The Fiske Model of Warfare', *Opns Res*. Vol. 10, No. 3, pp. 569-571.

95.* White, D.J., (1975) 'Decision Methodology', John Wiley and Sons Ltd., London.

96. Willis, R.F., (1982a) 'Alternative Concepts for the Design of Theories of Combat', RMCS Department of Management Sciences Working Paper OR/WP/25.

97. Willis, R.F., (1982b) 'Potential Uses for Historical Combat Data and Information', RMCS Department of Management Sciences Working Paper OR/WP/16.

98. Willis, R.F., (1982c) Collection of papers on deterministic and stochastic models, RMCS Department of Management Sciences Working Papers OR/WP/18, 19, 24, 26, 28, 32, 33.

99. Wolstenholme, E.F., (1982) 'System Dynamics in
     Perspective', *J. Opl Res. Soc.* Vol. 33, No. 6, pp. 547-556.

100. Wood, J.P., (1982) "Very Grave Suspicion", RUSI Journal,
     Vol. 127, No. 1, pp. 54-58.

GLOSSARY

Operational Analysis (OA): an aid to decision-making which
introduces a quantitative element, otherwise absent, by using
a scientific method.

Defence Operational Analysis (Defence OA): the branch of
operational analysis which deals with the problems of defence
(= Practice of Defence Studies + Analysis of Defence
Operations).

Study (with a capital S): a piece of analysis done in a
limited time in answer to a specific question and sponsored by
the Armed Services to assist the making of a decision.

Practice of Defence Studies: the carrying out of Studies by
teams of civilian analysts and military officers.

Analysis of Defence Operations: the analysis of defence
operations, outside the practice of defence Studies, without
reference to a need to make a specific operational or planning
decision.

Defence Operations: all the activities undertaken by the
Armed Services in peacetime, tension or war.

Science of Defence Operations: a permanent body of authentic
quantitative information on defence operations.

Combat Dynamics: the branch of defence operations concerned
with the particular activity of combat.

Science of Combat Dynamics: the branch of the science of
defence operations which describes what happens in combat
in defined circumstances.

Mathematical Modelling of Combat: the analysis of combat
through the use of symbolic and numerical mathematical models
(= Combat Theory + Combat Modelling).

Combat Theory: the branch of the mathematical modelling of
combat which is largely symbolic, usually manual, and involves
the solving of problems.

<u>Combat Modelling</u>: the branch of the mathematical modelling of combat which is essentially numerical, wholly computerised, and involves the running of models (that is, involves doing numerical experiments).

<u>Lanchester Theory</u> (of combat): the branch of combat theory which is based on the solutions of sets of differential and difference equations.

<u>Simulation Modelling</u> (of combat): the branch of combat modelling which is based on large, complex, mainly stochastic, simulation models.